Basic
Oxyacetylene
Welding

Basic Oxyacetylene Welding

FOURTH EDITION

IVAN H. GRIFFIN • EDWARD M. RODEN •
CHARLES W. BRIGGS

DELMAR PUBLISHERS INC.®

Cover photo by Larry Jeffus

Administrative editor: Mark Huth
Production editor: Eleanor Isenhart

For information, address Delmar Publishers Inc.,
2 Computer Drive West, Box 15-015,
Albany, New York 12212

COPYRIGHT © 1984
BY DELMAR PUBLISHERS INC.

All rights reserved. Certain portions of this work copyright © 1960, 1971, and 1977. No part of this work covered by the copyright hereon may be reproduced or used in any form or by any means — graphic, electronic, or mechanical, including photocopying, recording, taping, or information storage and retrieval systems — without written permission of the publisher.

Printed in the United States of America
Published simultaneously in Canada
by Nelson Canada,
A Division of International Thomson Limited

10 9 8 7 6

Library of Congress Cataloging in Publication Data
Griffin, Ivan H.
  Basic oxyacetylene welding.

  1 Oxyacetylene welding and cutting. I. Roden,
Edward M. II. Briggs, Charles W. III. Title.
TS228.G75 1984   671.5'2   83-72061
ISBN 0-8273-2137-6
ISBN 0-8273-2138-4 (instructor's guide)

CONTENTS

Preface		vii
Unit 1	The Oxyacetylene Welding Process	1
Unit 2	Oxygen and Acetylene Cylinders	4
Unit 3	Welding Gases	7
Unit 4	Oxygen and Acetylene Regulators	9
Unit 5	Types and Uses of Welding Torches	12
Unit 6	Welding Tips	14
Unit 7	The Oxyacetylene Welding Flame	17
Unit 8	Setting Up Equipment and Lighting the Torch	20
Unit 9	Flame Cutting	24
Unit 10	Straight Line Cutting	28
Unit 11	Bevel Cutting	31
Unit 12	Piercing and Hole Cutting	33
Unit 13	Welding Symbols	36
Unit 14	Running Beads and Observing Effects	41
Unit 15	Making Beads with Welding Rod	45
Unit 16	Tacking Light Steel Plate and Making Butt Welds	48
Unit 17	Flat Corner Welds	51
Unit 18	Lap Welds on Light Steel Plate	54
Unit 19	Tee or Fillet Welds on Light Steel Plate	56
Unit 20	Beads or Welds on Heavy Steel Plate	59
Unit 21	Manipulation of Welding Rod on Heavy Steel Plate	61
Unit 22	Butt Welds on Heavy Steel Plate	62
Unit 23	Lap Welds on Heavy Steel Plate	64
Unit 24	Fillet Welds on Heavy Steel Plate	66
Unit 25	Beveled Butt Weld on Heavy Steel Plate	68
Unit 26	Backhand Welding on Heavy Steel Plate	70
Unit 27	Backhand Welding of Beveled Butt Joints	72
Unit 28	Brazing with Bronze Rod	74
Unit 29	Running Beads with Bronze Rod	76
Unit 30	Square Butt Brazing on Light Steel Plate	78
Unit 31	Brazed Lap Joints	80
Unit 32	Brazed Tee Joints	82
Unit 33	Brazing Beveled Butt Joints on Heavy Steel Plate	84
Unit 34	Building-up on Cast Iron	86
Unit 35	Brazing Beveled Joints on Cast Iron	88
Unit 36	Silver Soldering Nonferrous Metals	90
Unit 37	Silver Soldering Ferrous and Nonferrous Metals	93
Index		97

CHARTS

Application Chart — Basic Oxyacetylene Welding ix

Contents

Chart 3-1 Some Comparative Temperatures	7
Chart 6-1 Material Thicknesses and Gas Pressures	14
Chart 6-2 Comparison Guide for Welding Tip Sizes	15
Chart 9-1 Comparison Guide for Cutting Tip Sizes	25
Chart 9-2 Gas Pressures and Kerf Widths for Cutting Various Thicknesses	26

PREFACE

The oxyacetylene flame is the oldest and still one of the most important sources of heat for welding metals and it is also frequently used for cutting metals. *Basic Oxyacetylene Welding* can be used to help the beginning welder develop skill in welding and cutting with oxyacetylene equipment.

This basic textbook acquaints the beginning welder with the use of the oxyacetylene flame in welding and cutting. Most of the units include step-by-step procedures for basic welding and cutting on steel plate or sheet metal. Each unit explores a new aspect of welding or cutting with the oxyacetylene flame. The introductory units present the fundamental information. After this introductory section, the emphasis on procedure helps in developing eye-hand coordination. This coordination is developed through experimentation and practice. The student can see the effect of varying the procedure manipulation of the equipment. Experimentation is encouraged. All of the units include review questions to check the students progress or to stimulate thought.

This revision of *Basic Oxyacetylene Welding* includes photographs of welds being made. The student can visualize what is actually taking place. All of the unit material, both new and old has been reviewed for readability and reliability. The addition of new illustrations and photographs helps to update the text. An index has been added to help the reader find specific information quickly and easily.

Because of its clear and readable format, *Basic Oxyacetylene Welding* has been popular in prevocational, industrial arts, adult education, and occupational education programs. It will provide an excellent foundation for the person who wants to enter the welding trade, or for those who will use oxyacetylene welding in another vocation.

ACKNOWLEDGMENTS

Appreciation is expressed to the following for their assistance in the development of this text:

- The American Welding Society for permission to use and adapt the Chart of Standard Welding Symbols; and the illustration of Common Faults that Occur in Hand Cutting.
- Air Reduction Sales Company (Airco) for permission to use and adapt certain illustrations.
- Linde Air Products Division of the Union Carbide Corporation for permission to use and adapt certain illustrations.
- Larry Jeffus, technical photographer and welding consultant.

APPLICATION CHART — BASIC OXYACETYLENE WELDING

Legend: R = Required, O = Optional, N = Not Necessary

Unit Number	Auto Mechanic	Boilermaker	Bricklayer	Carpenter	Electrician	Farm Equipment Repair	General Welding	Iron Worker (Ornamental)	Iron Worker (Structural)	Machinist	Plumber	Sheet Metal Worker	Steamfitter
1	R	R	R	R	R	R	R	R	R	R	R	R	R
2	R	R	R	R	R	R	R	R	R	R	R	R	R
3	R	R	R	R	R	R	R	R	R	R	R	R	R
4	R	R	R	R	R	R	R	R	R	R	R	R	R
5	R	R	R	R	R	R	R	R	R	R	R	R	R
6	R	R	R	R	R	R	R	R	R	R	R	R	R
7	R	R	R	R	R	R	R	R	R	R	R	R	R
8	R	R	R	R	R	R	R	R	R	R	R	R	R
9	R	R	R	R	R	R	R	R	R	R	R	R	R
10	N	R	O	N	N	R	R	O	O	O	R	N	R
11	N	R	O	N	N	R	R	O	O	O	R	N	R
12	N	R	O	N	N	R	R	O	O	O	R	N	R
13	R	R	R	R	R	R	R	R	R	R	R	R	R
14	R	R	R	R	R	R	R	R	R	R	R	R	R
15	R	R	R	R	R	R	R	R	R	R	R	R	R
16	N	O	O	O	N	R	R	R	O	R	R	R	R
17	N	O	O	O	N	R	R	R	O	R	R	R	R
18	N	O	O	O	N	R	R	R	O	R	R	R	R
19	N	O	O	O	N	R	R	R	O	R	R	R	R
20	O	R	O	N	N	R	R	O	O	O	R	O	R
21	O	R	O	N	N	R	R	O	O	O	R	O	R
22	O	R	O	N	N	R	R	O	O	O	R	O	R
23	O	R	O	N	N	R	R	O	O	O	R	O	R
24	O	R	O	N	N	R	R	O	O	O	R	O	R
25	N	N	O	N	N	R	R	O	O	R	R	N	O
26	N	N	O	N	N	R	R	O	O	R	R	N	O
27	N	N	O	N	N	R	R	O	O	R	R	N	O
28	N	N	O	N	N	R	R	O	O	R	R	N	O
29	N	N	O	N	N	R	R	O	O	R	R	N	O
30	O	N	O	N	N	R	R	O	O	R	R	N	O
31	O	N	O	N	N	R	R	O	O	R	R	N	N
32	O	N	O	N	N	R	R	O	O	R	R	N	N
33	O	N	O	N	R	R	R	O	O	R	R	N	N
34	N	N	O	N	R	R	R	O	O	R	R	N	N
35	R	R	R	R	R	R	R	R	R	R	R	R	R
36	R	R	R	R	R	R	R	R	R	R	R	R	R
37	R	R	R	R	R	R	R	R	R	R	R	R	R

Note: The above chart is in terms of suggested minimums only. The final choice of course content is a function of the individual instructor, often with the advice of an industry advisory committee.

Unit 1

THE OXYACETYLENE WELDING PROCESS

Gas welding is one of the three basic nonpressure processes of joining metals by *fusion* alone. The process of joining two pieces by partially melting their surfaces and allowing them to flow together is called fusion. The other two fusion processes are arc welding and thermit welding. Each of the three types has advantages and disadvantages.

In oxyacetylene welding, one of the gas welding processes, the metal is heated by the hot flame of a gas-fed torch. The metal melts and fuses together to produce the weld. In many cases, additional metal from a welding rod is melted into the joint which becomes as strong as the base metal.

EQUIPMENT AND HAZARDS

The basic equipment and materials for welding by this process are:

1. Oxygen and acetylene gas supplied from cylinders to provide the flame.
2. Regulators and valves to control the flow of the gases.
3. Gages to measure the pressure of the gases.
4. Hoses to carry the gases to the torch.
5. A torch to mix the gases and to provide a handle for directing the flame.
6. A tip for the torch to control the flame.

The above equipment is described in some detail in the next several units. A thorough understanding of the equipment is highly important so that welding may be done safely as well as efficiently. The hazards arising from lack of understanding and improper use of the equipment are:

1. Burns to the operator or nearby persons.
2. Fires in buildings or materials.
3. Explosions resulting in personal injury and property damage.
4. Damage to welding equipment.

ADVANTAGES AND DISADVANTAGES

Oxyacetylene welding, brazing, and soldering operations, which are carried out with similar equipment, have certain advantages and disadvantages.

1. Oxyacetylene welding is a process which can be applied to a wide variety of manufacturing and maintenance situations.
2. The equipment is portable.
3. The cost and maintenance of the welding equipment is low when compared to that of some other welding processes.
4. The cost of welding gases, supplies, and operator's time, depends on the material being joined and the size, shape, and position in which the weld must be made.

5. The rate of heating and cooling is relatively slow. In some cases, this is an advantage. In other cases where a rapid heating and cooling cycle is desirable, the oxyacetylene welding process is not suitable.
6. A skilled welder can control the amount of heat supplied to the joint being welded. This is always a distinct advantage.
7. The oxygen and nitrogen in the air must be kept from combining with the metal to form harmful oxides and nitrides.

In general, the oxyacetylene process can be used to advantage in the following situations:
- When the materials being joined are thin;
- When excessively high temperatures, or rapid heating and cooling of the work would produce unwanted or harmful changes in the metal.
- When extremely high temperatures would cause certain elements in the metal to escape into the atmosphere.

SAFETY

Many of the hazards in oxyacetylene welding can be minimized by careful consideration of the following points:

1. The welding flame and the sparks coming from the molten puddle can cause any flammable material to ignite on contact. Therefore,
   - Flame-resistant clothing must be worn by the welder and the welder's hair must be protected.
   - The welder should wear gloves designed for welding.
   - Welding and cutting should not be done near flammable materials such as wood, oil, waste or cleaning rags.
   - Never weld or cut on enclosed containers.
2. In addition to the risk of eye injury from flying sparks, the eyes may be strained if the intense light is not filtered out by proper lenses. Therefore,
   - The eyes should be protected at all times by approved safety glasses and the proper shield. A number 5 lens is the standard density.
   - Sunglasses are not adequate for this purpose.
3. Fluxes used in certain welding and brazing operations produce fumes which are irritating to the eyes, nose, throat, and lungs. Likewise, the fumes produced by overheating lead, zinc, and cadmium are a definite health hazard when inhaled even in small quantities. The oxides produced by these elements are poisonous. Therefore,
   - Welding should be done in a well-ventilated area.
   - The operator should not expose others to fumes produced by welding.

REVIEW QUESTIONS

1. What two gases in the air are harmful when combined with metal?

The Oxyacetylene Welding Process

2. What features of the oxyacetylene welding process make it useful in a wide variety of jobs?

3. What are four distinct hazards that must be guarded against when oxyacetylene welding?

4. Name three metals that can produce toxic fumes when overheated with a welding torch.

5. What is the function of a regulator?

Unit 2

OXYGEN AND ACETYLENE CYLINDERS

Hazards are always present when gases are compressed, stored, transported, and used under very high pressures. Oxygen and acetylene are delivered to the user under high pressure in steel cylinders. These cylinders are made to rigid specifications.

Cylinders are built, filled and transported according to interstate commerce commission regulations. Before transporting, all cylinders must be secured in an upright position and cylinder caps must be in place.

A simple demonstration of the effects of compressing gas can be shown with an ordinary toy balloon. When the balloon is blown up and held tightly at the neck so the air cannot escape, it resembles the compressed gas cylinder. What happens if the balloon is punctured, heated, or suddenly released? The explosive burst when punctured or heated, or the sudden flight of the balloon when released shows that compressed gas, even the small amount in a toy balloon, has considerable force.

OXYGEN CYLINDERS

The most common size of oxygen cylinder, when fully charged with gas, contains 244 cubic feet of oxygen. This oxygen is at a pressure of 2,200 pounds per square inch when the temperature is 70 degrees F. (21 degrees C).

The steel walls of these cylinders are only slightly more than one-quarter inch thick, .260 inch. Dropping such a cylinder, hitting it with heavy or sharp tools, or striking an electric arc on it can cause the cylinder to explode with enough force to cause serious injury and death.

The general size and shape of an oxygen cylinder is indicated in figure 2-1. As a safety precaution, the cylinder valve is protected by a removable steel cap. This cap must be on the cylinder at all times when it is being stored or transported. The cylinder valve should always be closed when not in use, even when the cylinder is empty.

The oxygen cylinder valve is designed to handle the highly compressed oxygen gas safely. The essential parts of the valve are shown in figure 2-2. The threads on the nozzle must be protected at all times.

The *bursting disc* and *safety cap* are designed to allow the gas in the cylinder to escape if the cylinder is subjected to undue heat and the pressure in the tank begins to rise.

The double-seating valve is designed to seal off any oxygen that might leak around the valve stem. When the valve is fully open there is no leakage.

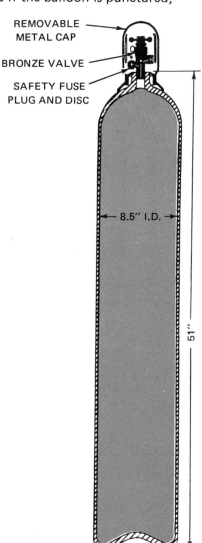

Fig. 2-1 Oxygen cylinder

Oxygen and Acetylene Cylinders

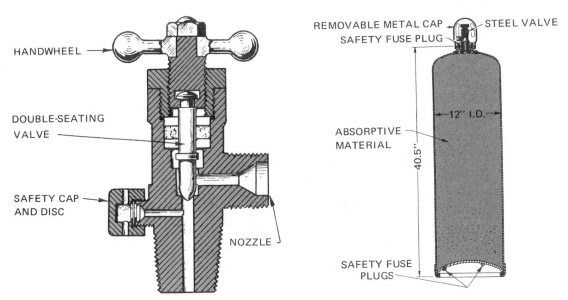

Fig. 2-2 Oxygen-cylinder valve   Fig. 2-3 Acetylene cylinder

ACETYLENE CYLINDERS

The acetylene cylinder is a welded steel tube. It is filled with a spongy material such as balsa wood or some other absorptive material which is saturated with a chemical solvent called acetone. Acetone absorbs acetylene gas in much the same manner as water absorbs ammonia gas to produce common household ammonia.

Some cylinders are equipped with a valve which can only be opened with a special wrench. Safety regulations require this type of valve on all containers carrying flammable, explosive, or toxic gases. The wrench must be in place whenever the cylinder is in use. Acetylene cylinders are also equipped with a number of *fusible plugs* designed to melt at 220 degrees F. These melt and release the pressure in the event the cylinder is exposed to excessive heat.

Figure 2-3 is a cross section of a common acetylene cylinder. The construction details may vary from one manufacturer to another, but all acetylene cylinders are made to very rigid specifications.

Acetylene cylinders are usually charged to a pressure of 250 pounds per square inch; the large size contains about 275 cubic feet. The steel walls of these cylinders are .175 inch thick. The precautions set forth for oxygen cylinders should be observed with acetylene cylinders. Escaping acetylene mixed with air forms a highly explosive mixture. Never operate acetylene cylinders in a horizontal position.

REVIEW QUESTIONS

1. What prevents unauthorized persons from opening acetylene valves?

Basic Oxyacetylene Welding

2. Why must an oxygen cylinder valve be fully opened?

3. What happens to the pressure in a cylinder as the temperature is raised?

   What happens when the temperature is lowered?

4. Suppose the protective cap is left off a fully-charged oxygen cylinder and an accident causes the valve to be broken off:

   a. What is the force of the gas per square inch?

   b. Is this enough force to cause the cylinder to move?

5. What is the function of the cylinder valve?

Unit 3
WELDING GASES

OXYGEN

Flame is produced by combining oxygen with other materials. When the air we breathe, which is only one-fifth oxygen, combines with other elements to produce a flame, this flame is low in temperature and the rate of burning is rather slow.

However, if pure oxygen is substituted for air, the burning is much more rapid and the temperature is much higher. Oil in the regulators, hoses, torches, or even in open air burns with explosive rapidity when exposed to pure oxygen.

CAUTION: Oxygen must never be allowed to come in contact with any flammable material without proper controls and equipment. The use of oxygen to blow dust and dirt from working surfaces or from a worker's hair or clothing is extremely dangerous. Never use oxygen for compressed air purposes.

Most of the oxygen produced commercially in the United States is made by liquefying air and then recovering the pure oxygen. The oxygen thus produced is of such high purity that it can be used not only to produce the most efficient flame for welding and flame cutting operations, but also for medical purposes.

Oxygen is a colorless, odorless, tasteless gas which is slightly heavier than air. The weight of 12.07 cubic feet of oxygen at atmospheric pressure and 70 degrees F. is one pound.

CHART 3-1

SOME COMPARATIVE TEMPERATURES		
Electric Arc	11,000 F	6,090 C
Surface of Sun	10,832 F	6,000 C
Oxyacetylene Flame	5,900 F	3,260 C
Oxyhydrogen Flame	5,252 F	2,900 C
Interior of Internal Combustion Engine	3,272 F	1,800 C
Copper Melts	1,976 F	1,080 C
Magnesium Melts	1,204 F	651 C
Water Boils	212 F	100 C
Ice Melts	32 F	0 C
Liquid Air Boils	- 292 F	- 180 C
Liquid Helium Boils	- 452 F	- 269 C
Absolute Zero	- 459.4 F	- 273 C

ACETYLENE

Acetylene gas is a chemical compound composed of carbon and hydrogen. It combines with oxygen to produce the hottest gas flame known. Unfortunately, acetylene is an unstable compound and must be handled properly to avoid explosions.

Unstable acetylene gas tends to break down chemically when under a pressure greater than 15 pounds per square inch. This chemical breakdown produces great amounts of heat; the resulting high pressure develops so rapidly that a violent explosion may result.

Acetylene gas which is dissolved in acetone does not tend to break down chemically and can be used with complete safety. However, any attempt to compress acetylene in a free state in hoses, pipes, or cylinders at a pressure greater than 15 pounds per square inch can be very dangerous.

Acetylene is produced by dissolving calcium carbide in water. This process should be carried out only in approved generators. One pound of calcium carbide produces 4.5 cubic feet of acetylene gas. Acetylene is made up of two atoms of carbon and two atoms of

Basic Oxyacetylene Welding

hydrogen. It has a distinctive odor. The weight of 14.5 cubic feet is one pound. The amount dissolved in an acetylene cylinder is determined by weighing the cylinder and contents, subtracting the weight of the empty cylinder, and multiplying the remainder, which is the weight of the gas, by 14.5. The empty cylinder weight is always stamped into the cylinder.

REVIEW QUESTIONS

1. Why is it dangerous to place calcium carbide and water in a closed container and generate acetylene gas?

2. What is the probable effect if oil or grease is allowed to come in contact with oxygen in the regulators or cylinders?

3. What element must always be present if a flame is to be produced and maintained?

4. At what pressure will acetylene gas become unstable in a free state?

5. Can oxygen be referred to as air? Why?

Unit 4

OXYGEN AND ACETYLENE REGULATORS

REGULATORS

Oxygen and acetylene *regulators* reduce the high cylinder pressures, safely and efficiently, to usable working pressures. Regulators also maintain these pressures within very close limits under varying conditions of demand.

Figure 4-1 shows the relatively simple operation of a regulator. The pressure in the hoses is controlled by applying pressure to the spring through an adjusting screw. The spring applies pressure to a flexible rubber diaphragm which is connected to the high-pressure valve. The gas from the cylinder flowing through this valve builds up pressure behind the diaphragm. When this pressure equals the pressure of the spring, the valve closes and shuts off the flow of gas to the diaphragm area. When the pressure in this area is reduced by drawing gas from the regulator to the torch the spring opens the valve again.

Regulators are made to rigid specifications from the finest of materials, and are equipped with safety devices to prevent injury to the operator or the equipment. All regulators are equipped with ball-check safety valves or bursting discs to prevent pressure buildup within the regulator, hoses, or torch.

GAGES

Most regulators are equipped with gages which indicate the amount of pressure in the cylinder and the working pressure in the hoses and torch.

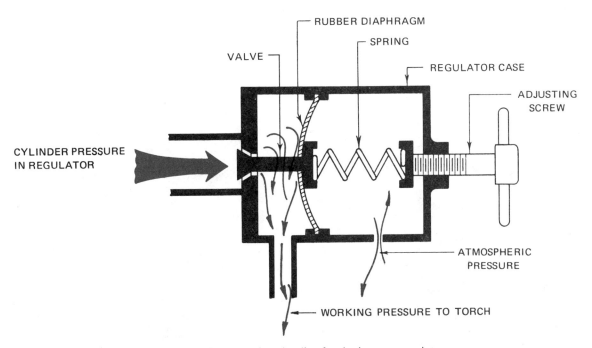

Fig. 4-1 Construction details of a single-stage regulator

Basic Oxyacetylene Welding

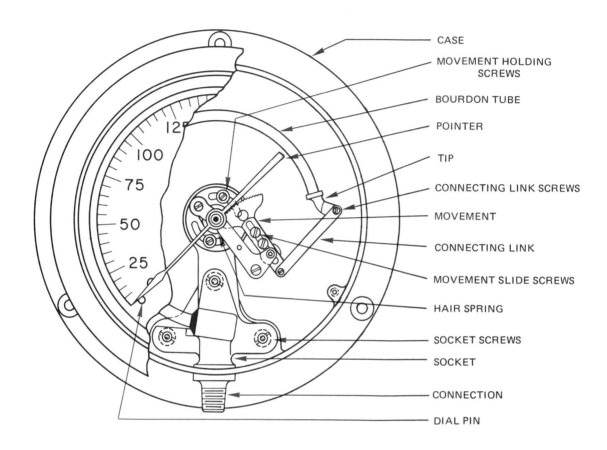

Fig. 4-2 Construction details of a gage

Fig. 4-3 Oxygen regulator

Fig. 4-4 Acetylene regulator

Oxygen and Acetylene Regulators

These gages have very thin backs which open to release the pressure if the Bourdon tube ruptures. This tube is essential to the operation of each gage. If this precaution was not taken, excessive pressure in the gage case could cause the glass front of the gage to explode and injure the operator.

Since gages frequently get out of calibration, they are only indicators of cylinder and working pressures. Regulators work regardless of the accuracy of the gages.

Unit 8 describes the procedure that should be followed to insure safety and efficiency when adjusting the regulators, regardless of the pressures indicated on the gages.

REVIEW QUESTIONS

1. What two purposes do oxygen and acetylene regulators serve?

2. What purpose do gages on regulators serve?

3. Should a regulator-adjusting screw be turned all the way in when the regulator is to be turned off?

4. What safety devices guard against excessive pressure in regulator cases and in hoses?

5. Can a regulator be accurate if the gage is damaged?

Unit 5

TYPES AND USES OF WELDING TORCHES

The body of the welding torch serves as a handle so the operator can hold and direct the flame. Beyond the handle, the torch is equipped with a means of attaching the mixing head and welding tip, figure 5-1.

The accurately sized holes in welding and cutting tips are called *orifices*. The purpose of the *mixing head* is to combine the two welding gases into a usable form. The only mixed oxygen and acetylene is that amount contained in the space from the mixing head to the tip orifice. In most cases, it represents a very small portion of a cubic inch. This keeps the amount of this highly explosive mixture within safe limits. Any attempt to mix greater amounts may result in violent explosions.

Two types of torches are in common use. In the *injector-type torch,* the acetylene at low pressure is carried through the torch and tip by the force of the higher oxygen pressure through a venturi-type device, shown in figure 5-2. The mixing head and injector are usually a part of the tip which the operator changes according to the size needed.

In the *medium-pressure torch,* figure 5-3, both gases are delivered through the torch to the tip at equal pressures. In this type of torch, the mixer or mixing head is usually a separate piece into which a variety of tips may be fitted.

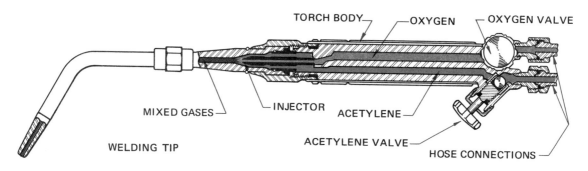

Fig. 5-1 Welding torch

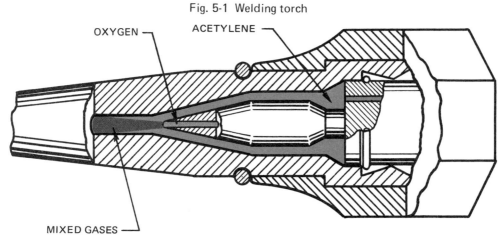

Fig. 5-2 Injector-type mixer

Types and Uses of Welding Torches

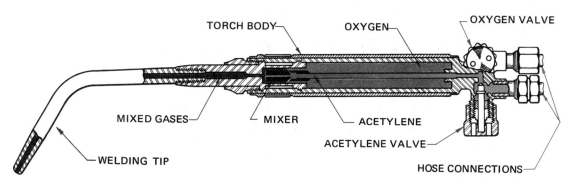

Fig. 5-3 Medium-pressure torch

Fig. 5-4 Welding torch

All types of torches are equipped with a pair of needle valves to turn the welding gases on and off, and to make small pressure adjustments.

REVIEW QUESTIONS

1. What is the chief difference between the two types of torches?

2. a. What is the basic purpose of torch needle valves?

   b. What secondary purpose do they serve?

3. From the construction details indicated in the text, which type of torch is probably more costly?

4. What is the proper name for the holes in welding and cutting tips?

5. How many valves do most torches have?

Unit 6

WELDING TIPS

The purpose of all welding tips is to provide a safe, convenient method of varying the amount of heat supplied to the weld. They also provide a convenient method of directing the flame and heat to the exact place the operator chooses.

SELECTION OF SIZES

To provide for different amounts of heat, welding tips are made in various sizes. The size is determined by the size of the orifice in the tip. As the orifice size increases, greater amounts of the welding gases pass through and are burned to supply a greater amount of heat. However, the temperature of the neutral oxyacetylene flame always remains at 5,900 degrees F., regardless of the quantity of heat provided.

The choice of the proper tip size is very important to good welding. All manufacturers of welding torches supply a chart of recommended sizes for various thicknesses of metal. They also recommend oxygen and acetylene pressures for various types and sizes of tips, chart 6-1.

CARE OF TIPS

All welding tips are made of copper and may be damaged by careless handling. Dropping, prying, or hammering the tips on the work may make them unfit for further use. It is im-

CHART 6-1

Plate Thicknesses		Gas Pressures in P.S.I.			
		Injector-type Torch		Equal-pressure Torch	
Gage	Inches	Acetylene	Oxygen	Acetylene	Oxygen
32	.010	5	5 - 7	1	1
28	.016	5	7 - 8	1	1
26	.019	5	7 - 10	1	1
22	$\frac{1}{32}$	5	7 - 18	2	2
16	$\frac{1}{16}$	5	8 - 20	3	3
13	$\frac{3}{32}$	5	15 - 20	4	4
11	$\frac{1}{8}$	5	12 - 24	4	4
8	$\frac{3}{16}$	5	16 - 25	5	5
	$\frac{1}{4}$	5	20 - 29	6	6
	$\frac{3}{8}$	5	24 - 33	7	7
	$\frac{1}{2}$	5	29 - 34	8	8
	$\frac{5}{8}$	5	30 - 40	9	9
	$\frac{3}{4}$	5	30 - 40	10	10
	1	5	30 - 42	12	12

Welding Tips

CHART 6-2
COMPARISON GUIDE FOR WELDING TIP SIZES

00 — Tip Size
000 — Orifice Size

Welding Tips

Tip Name	.015"/28 Ga.	.030"/22 Ga.	.048"/18 Ga.	.060"/16 Ga.	3/32"/13 Ga.	1/8"/11 Ga.	5/32"/8 Ga.	3/16"/6 Ga.	1/4"	5/16"	7/16"	5/8"	3/4"	1"	1" & Over
Airco	00	0	0	1	2	3	3	4	5	6	7	8	9	10	10
	.020	.025	.025	.031	.038	.047	.047	.055	.067	.076	.086	.098	.110	.128	.128
Dockson	1	1	2	2	3	4	5	5	6	7	9	11	12	14	15
	.016	.016	.024	.024	.033	.042	.052	.052	.063	.076	.096	.116	.128	.147	.154
Harris	0	1	2	3	4	5	5	6	7	8	9	10	13	15	19
	.021	.026	.035	.042	.052	.059	.059	.067	.076	.082	.089	.098	.110	.120	.144
KG	00	0	0	1	2	3	4	4	5	6	7	8	9	10	10
	.020	.025	.025	.031	.038	.046	.055	.055	.067	.076	.086	.098	.110	.128	.128
Meco	00	0	1	2	3	4	4	5	5	6	7	8	9	10	11
	.013	.020	.028	.035	.043	.052	.052	.063	.063	.078	.093	.110	.128	.140	.157
Oxweld	—	4	4	6	6	9	15	15	15	30	30	55	55	70	100
	—	.036	.036	.042	.042	.052	.059	.059	.059	.082	.082	.113	.113	.128	.152
Purox	2	4	6	6	9	15	15	15	30	30	30	55	55	70	100
	.022	.036	.042	.042	.052	.059	.059	.059	.082	.082	.082	.113	.113	.128	.152
Rego	79	76	72	68	62	58	55	53	50	46	42	36	31	25	20
	.014	.020	.025	.031	.038	.042	.052	.059	.070	.081	.093	.106	.120	.149	.161
Smith	000	00	1	2	4	5	6	7	8	9	9	11	11	12	13
	.016	.020	.026	.029	.037	.043	.046	.055	.063	.073	.073	.098	.098	.111	.128
Victor	000	000	00	0	1	2	3	3	4	5	5	6	7	7	7
	.021	.021	.028	.035	.040	.046	.059	.059	.073	.089	.089	.106	.128	.128	.128

Data is subject to change by manufacturer.

When cleaning tips, it is recommended that the tip cleaner be one size smaller than the orifice size.

Basic Oxyacetylene Welding

portant to clean the tip orifice with the proper tip drill. The use of an incorrect drill or procedure can ruin a tip.

REVIEW QUESTIONS

1. What size Airco tip should be used for welding 3/16" steel?

2. What size Purox tip should be used for welding 16 gage steel?

3. What size Victor tip should be used for welding 18 gage steel?

4. Welding tips are made of what type of material?

5. What is the decimal equivalent for the thickness of 16 gage steel?

6. What are the holes in welding tips called?

Unit 7

THE OXYACETYLENE WELDING FLAME

The flame is the actual tool of oxyacetylene welding. All of the welding equipment merely serves to maintain and control the flame.

The flame must be of the proper size, shape, and condition in order to operate with maximum efficiency. The oxyacetylene flame differs from most other types of tools in that it is not ready-made. The operator must produce the proper flame each time the torch is lit.

Once the operator masters the adjustment of the flame, the operator's ability as a welder increases in direct proportion to the amount of practice the operator has.

TYPES OF FLAMES

The oxyacetylene flame can be adjusted to produce three distinctly different types of flame. Each of these types has a very marked effect on the metal being fused or welded. In the order of their general use, the flames are *neutral, carburizing,* and *oxidizing*. Figure 7-1 illustrates their shapes and characteristics.

The *neutral* flame, figure 7-2, is one in which equal amounts of oxygen and acetylene combine in the inner cone to produce a flame with a temperature of 5,900 degrees F. The inner cone is light blue in color. It is surrounded by an outer flame envelope, produced by the combination of oxygen in the air and superheated carbon monoxide and hydrogen gases from the inner cone. This envelope is usually a much darker blue than the inner cone. The advantage of the neutral flame is that it adds nothing to the metal and takes nothing away. Once the metal has been fused, it is chemically the same as before welding.

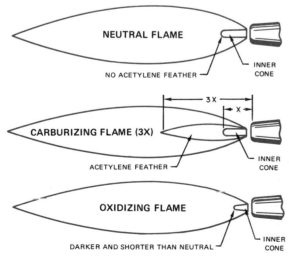

Fig. 7-1 Types of flames

Fig. 7-2 Neutral flame (balanced oxygen and acetylene) (Reprinted from Jeffus & Johnson, *Welding: Principles & Applications,* Figure 5-8. © 1984 by Delmar Publishers Inc.)

Basic Oxyacetylene Welding

Fig. 7-3 Carburizing flame (excessive acetylene) (Reprinted from Jeffus & Johnson, *Welding: Principles & Applications,* Figure 5-7)

The *carburizing* flame (reducing flame), figure 7-3, is indicated by streamers of excess acetylene from the inner cone. These streamers are usually called *feathers* of acetylene, or simply the *acetylene feather.* The feather length depends on the amount of excess acetylene. The outer flame envelope is longer than that of the neutral flame and is usually much brighter in color. This excess acetylene is very rich in carbon. When carbon is applied to red-hot or molten metal, it tends to combine with steel and iron to produce the very hard, brittle substance known as iron carbide. This chemical change leaves the metal in the weld unfit for many applications in which the weld may need to be bent or stretched. While this type of flame does have its uses, it should be avoided when fusion welding those metals which tend to absorb carbon.

The carburizing flame in figures 7-1 and 7-3 shows the relation of the acetylene streamers to the inner cone. Job conditions sometime require an excess of acetylene in terms of the length of the inner cone.

The *oxidizing* flame, figure 7-4, which has an excess of oxygen, is probably the least used of any of the three flames. In appearance, the inner cone is shorter, much bluer in color, and usually more pointed than a neutral flame. The outer flame envelope is much shorter and tends to fan out at the end. The neutral and carburizing envelopes tend to come to a sharp point. The excess oxygen in the flame causes the temperature to rise as high as 6,300 degrees F. This temperature would be an advantage if it were not for the fact that the excess oxygen, especially at high temperatures, tends to combine with many metals to form hard, brittle, low-strength oxides. For this reason, even slightly oxidizing flames should be avoided in welding.

Fig. 7-4 Oxidizing flame (excessive oxygen) (Reprinted from Jeffus & Johnson, *Welding: Principles & Applications,* Figure 5-9)

The Oxyacetylene Welding Flame

REVIEW QUESTIONS

1. What chemical change takes place when a **carburizing** flame is used to weld steel?

2. What substance is produced when an oxidizing flame is used to weld steel?

3. Make a labeled and dimensioned sketch of a 2X flame. What are the significant parts?

4. Make a labeled and dimensioned sketch of a 3 1/2X flame. What are the significant parts?

5. What is the temperature of a neutral flame?

Unit 8

SETTING UP EQUIPMENT AND LIGHTING THE TORCH

Oxyacetylene welding equipment, figure 8-1, must be set up frequently and it must be done efficiently. Since hazards are present, each step must be performed correctly. The proper sequence must be followed to insure maximum safety to personnel and equipment.

The cylinder caps are removed and put in their proper place. The cylinders should be fastened to a wall or other structure with chains, straps or bars, to prevent them from being tipped over. To use oxygen and acetylene cylinders and equipment without this safety precaution is to invite damage to the equipment and injury to the operator.

Procedure

1. Aim the cylinder nozzle so it does not blow toward anyone. Crack the valve on each cylinder by opening the valve and closing it quickly. This blows any dust or other foreign material from the nozzle.
2. Attach the regulators to the cylinder nozzles.

    Note: All oxygen regulators in commercial use have a standard *right-hand* thread and fit all standard oxygen cylinders. Acetylene regulators may have *right- or left-hand* threads and may have either a male or female connection, depending on the company supplying the gas. Adapters of various types may be needed to fit the existing regulators to different acetylene cylinders.

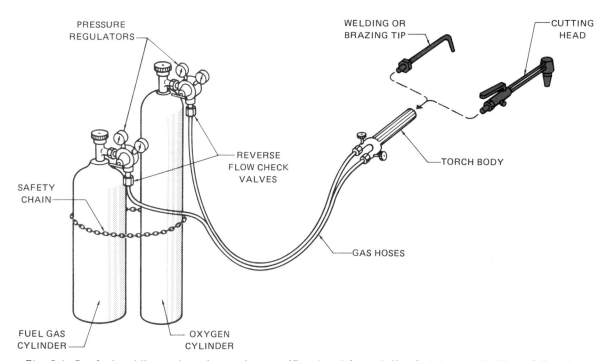

Fig. 8-1 Oxyfuel welding and cutting equipment (Reprinted from Jeffus & Johnson, *Welding: Principles & Applications,* Figure 1-4)

Setting Up Equipment and Lighting the Torch

3. Attach the hoses to the regulators.

   Note: Oxygen hoses are green or black; acetylene hoses are red. All oxygen hose connections have *right-hand* threads; all acetylene hose connections have *left-hand* threads. The acetylene hose connection nuts are distinguished from the oxygen nuts by a *groove* machined around the center of the nut, figure 8-2. Check valves should be placed between the hoses and torch to prevent gases and fire from backing into hoses and regulators.

4. Attach the torch to the other end of the hoses noting that while the hose connections may be a different size at the torch than at the regulators, they still have right- and left-hand threads.

   Note: Use only the wrenches provided for attaching hoses and regulators. These wrenches are designed to give the proper leverage to tighten the joints without putting undue strain on the equipment. If the joints cannot be properly tightened, something is wrong. Never use sealer on threads.

5. Select the proper tip and mixing head and attach it to the torch. Position the tip so that the needle valves are on the side or bottom of the torch when the tip is in the proper welding position.

6. Back off the regulator screws on both units until the screws turn freely. This is necessary to eliminate a sudden surge of excessive pressure on the working side of the regulator when the cylinder is turned on.

7. Be sure both torch needle valves are turned off (clockwise). This is an added safety precaution to make sure excessive pressure cannot be backed through the mixing head and into the opposite hose.

8. Open the acetylene cylinder valve 1/4 to 1/2 turn. Open the oxygen cylinder valve all the way.

9. Open the acetylene needle valve one full turn. Turn the adjusting screw on the acetylene regulator clockwise until gas comes from the tip. Light this gas with a sparklighter.

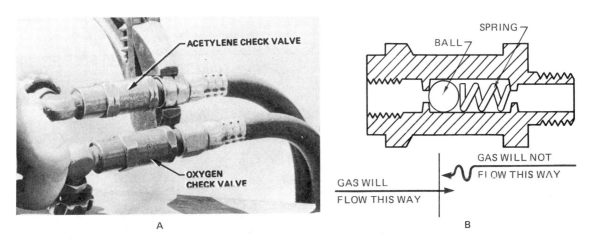

Fig. 8-2 Oxygen and acetylene hose connections showing check valves (Figure 8-2B is adapted from Jeffus, *Safety for Welders,* Figure 7-1. © 1980 by Delmar Publishers Inc.)

Basic Oxyacetylene Welding

10. Adjust the regulator screw until there is a gap of about 1/4 inch between the tip and the flame. This is the proper pressure for the size of tip being used regardless of the gage pressure shown on the working pressure gage.

11. Open the oxygen needle valve on the torch one full turn. Turn the oxygen regulator adjusting screw clockwise until the flame changes appearance as oxygen is mixed with the acetylene.

12. Continue to turn the adjusting screw until the feather of acetylene just disappears into the end of the inner cone. This produces a neutral flame which is used in most welding.

This procedure for adjusting the oxyacetylene flame is the safest method of insuring the proper working pressures in both hoses and tip. Working pressure gages are delicate and easily get out of calibration. If this happens, excessive pressure can be built up in the hoses before it is discovered. However, if the pressures are adapted to the flame as indicated, there are equal pressures in both hoses which eliminates the possibility of backing gas from one hose to the other to form an explosive mixture. With the regulators properly adjusted, minor flame adjustments are made with the torch needle valves.

When the welding or cutting operation is finished, close the torch acetylene valve first, then the torch oxygen valve.

To shut down the equipment for an extended period of time, such as overnight, it should be purged. Use the following procedure:

1. Close the oxygen cylinder valve.
2. Open the torch oxygen valve to release all pressure from the hose and regulator.
3. Turn out the pressure adjusting screw of the oxygen regulator.
4. Close the torch oxygen valve.
5. Follow the same sequence for purging acetylene.

REVIEW QUESTIONS

1. What are the steps necessary to turn on the welding gases properly and safely and adjust them to a suitable flame?

2. What steps are necessary to assemble an oxyacetylene outfit for welding?

Setting Up Equipment and Lighting the Torch

3. Why should the needle valves on the torch be turned off at a particular step in the sequence rather than at some other time?

4. Could excessive oxygen pressure backed through the torch cause an explosion in the acetylene hose without outside ignition? Explain. (Refer to Unit 3 — Acetylene Gas.)

5. How is a left-hand nut different from a right-hand nut in appearance?

Unit 9

FLAME CUTTING

One of the fastest ways of cutting mild steel is by the use of the oxyacetylene torch. Other advantages of this cutting method are:

1. A relatively smooth cut is produced.
2. Very thick steel can be cut.
3. The equipment is portable.
4. Underwater cutting is possible with some adaptations.
5. The equipment lends itself to automatic processes in manufacturing.

The terms "cutting" and "burning" are used interchangeably to describe this process.

THE CUTTING PROCESS

Oxyacetylene flame cutting is actually a burning (rapid-oxidation) process in which the metal to be cut is heated on the surface to the kindling temperature of steel (1,600–1,800 degrees F.). A small stream of pure oxygen is then directed at the work. The oxygen causes the metal to ignite and burn to produce more heat. This additional heat causes the nearby metal to burn so that the process is continuous once it has started.

Only those ferrous metals which oxidize rapidly can be flame-cut. These metals include all the straight carbon steels and many of the alloys. Stainless steels and most of the so-called high-speed steels cannot be flame-cut.

EQUIPMENT

Cutting is performed with a manual torch, figure 9-1 and with different machine torches figure 9-2 and figure 9-3. Tips for these torches are interchangeable so that they can be adapted to cut a wide variety of metal thicknesses. The torches and tips are constructed so that they can preheat the work to the kindling temperature. The manual torch also includes a lever for starting and stopping the stream of high-pressure cutting oxygen as required.

Most cutting tips are made of copper which is soft. If the tip of a manual torch is used as a hammer, lever or crowbar, permanent damage is done.

COMPARISON CHARTS

Because cutting torch tips are interchangeable, chart 9-1 may be used for the torch tips of all major manufacturers.

Fig. 9-1 Combination cutting torch

Flame Cutting

Fig. 9-2 Machine cutting torch

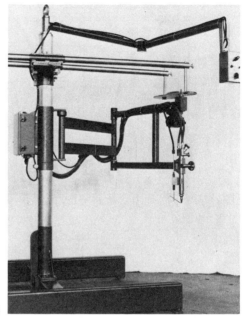

Fig. 9-3 Shape cutting machine

CHART 9-1

COMPARISON GUIDE FOR CUTTING TIP SIZES																
00 — Tip Size 000 — Orifice Size						Cutting Tips										
	Material Thickness															
Tip Name	1/8"	1/4"	3/8"	1/2"	3/4"	1"	1 1/2"	2"	3"	4"	5"	6"	7"	8"	9"	10"
Airco	00	0	0	1	2	2	3	3	4	5	5	6	6	7	7	8
	.031	.038	.038	.046	.055	.055	.063	.063	.073	.082	.082	.096	.096	.111	.111	.128
Dockson	0	1	1	1	2	2	3	3	4	4	5	5	6	6	7	7
	.035	.039	.039	.039	.052	.052	.073	.073	.089	.089	.106	.106	.116	.116	.140	.140
Harris	000	00	00	0	1	1	2	2	3	4	4	4	5	5	6	6
	.031	.036	.036	.040	.046	.046	.063	.063	.076	.093	.093	.093	.110	.110	.128	.128
KG	00	0	1	1	1	2	3	3	5	5	6	6	6	7	7	7
	.031	.038	.046	.046	.046	.055	.063	.063	.082	.082	.098	.098	.098	.110	.110	.110
Meco	00	0	0	0	1	1	2	2	3	4	5	5	6	6	6	7
	.032	.032	.032	.032	.043	.043	.055	.055	.067	.078	.093	.093	.110	.110	.110	.128
Oxweld	2	3	3	4	6	6	6	8	8	8	8	10	10	10	10	10
	.020	.031	.031	.040	.059	.059	.059	.081	.081	.081	.081	.100	.100	.100	.100	.100
Purox	3	3	4	4	5	5	5	7	7	7	9	9	11	11	11	11
	.031	.031	.040	.040	.052	.052	.052	.070	.070	.070	.089	.089	.110	.110	.110	.110
Rego	68	68	62	62	56	56	53	53	51	46	42	42	35	35	35	30
	.031	.031	.038	.038	.046	.046	.059	.059	.067	.081	.093	.093	.110	.110	.110	.128
Smith	000	0	0	1	2	2	3	3	4	4	5	5	5	5	6	6
	.025	.038	.038	.046	.055	.055	.070	.070	.082	.082	.100	.100	.100	.100	.120	.120
Victor	000	00	0	0	1	2	2	3	4	5	5	6	6	6	7	7
	.026	.032	.040	.040	.046	.059	.059	.070	.082	.100	.100	.120	.120	.120	.140	.140

Data is subject to change by manufacturer.

When cleaning tips, it is recommended that the tip cleaner be one size smaller than the orifice size.

Basic Oxyacetylene Welding

CHART 9-2

GAS PRESSURES AND KERF WIDTHS FOR CUTTING VARIOUS THICKNESSES			
Material Thickness (Inches)	Approximate Gage Pressure Oxygen (P.S.I.)	Approximate Gage Pressure Acetylene (P.S.I.)	Approximate Width of Kerf (Inches)
1/8	20 – 30	3 – 5	.075
1/4	20 – 30	3 – 5	.075
3/8	30 – 40	4 – 5	.095
1/2	30 – 50	4 – 5	.095
3/4	40 – 60	5	.110
1	40 – 60	5 – 7	.110
1 1/2	40 – 75	5 – 7	.130
2	40 – 75	5 – 8	.130

HAZARDS

The operator must protect his eyes at all times with goggles fitted with proper lenses, usually shade 5 or 6. Gauntlet-type gloves and any other equipment necessary to give protection from the molten iron oxide, should be worn.

Since the high-pressure stream of cutting oxygen can throw small bits of molten oxide, at a temperature of 3,000 degrees F. and up, for distances of 50-60 feet, the operator should check before starting the burning operation to be sure that all flammable and explosive materials have been removed to a safe place.

The operator should insure that all personnel in the area are warned of the shower of molten metal that will occur so that they may take the necessary precautions.

The International Acetylene Association and the Underwriters' Laboratories recommend that an additional workman with fire-fighting equipment be assigned to each unit during cutting and for 2 hours after completion of cutting to guard against fires.

REVIEW QUESTIONS

1. What are the limitations of flame cutting?

Flame Cutting

2. What chemical change takes place during the burning process?

3. What special safety precautions must be taken?

4. How wide must a piece of one-inch steel plate be so that 10 strips each 3 inches wide can be cut from it? Make the proper allowance for the width of the kerf (cut width) from chart 9-2.

5. Using chart 9-1, determine what size tip should be used with a Purox torch to cut 1/2" plate.

6. How far can molten particles be thrown from an oxyacetylene cutting torch?

Unit 10

STRAIGHT LINE CUTTING

Several things affect the speed, smoothness, and general quality of a cut made by an oxyacetylene flame. This unit provides practice in changing these variables to determine the best methods for flame cutting.

The actual cutting process demonstrates the danger of personal burns and fires which might cause property damage.

Materials

1/4-inch or 3/8-inch thick steel plate, approximately 4 in. X 10 in.
Cutting tip, see chart 9-1 for size.

Procedure

1. Draw a series of straight parallel lines about 2 inches apart on the plate. Use soapstone for marking so that the lines show up when the cutting goggles are being worn.

2. Light and adjust the preheating flame to neutral using the data supplied in chart 9-2.

3. Start the cut by holding the tip over the edge of the metal so that the vertical centerline of the tip is square with the work and in line with the edge of the plate. The tip is positioned in the torch as indicated in figure 10-1.

4. When the edge of the work becomes bright red, figure 10-2, turn the cutting oxygen on with the lever. Note that the oxygen makes a cut through the plate at the same angle that the centerline of the tip makes with the work.

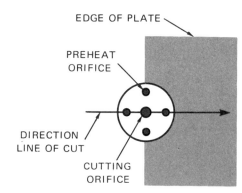

Note: Tips may have more than 4 preheat orifices.

Fig. 10-1 Tip alignment for square cuts

A

B

Fig. 10-2 When the steel is cherry red, depress the cutting lever. (Figure 10-2B is reprinted from Jeffus & Johnson, *Welding: Principles & Applications,* Figure 8-38A)

Straight Line Cutting

5. Continue the cut, making sure that the tip is square with the work. Observe that when the rate of travel is right, the slag or iron oxide coming from the cut makes a sound like cloth being torn. The tearing sound serves as a guide to the correct rate of travel in most manual flame-cutting operations.
6. Finish the cut and check the flame-cut edge for smoothness, straightness, and amount of slag on the bottom edge of the cut surfaces.
7. Make more cuts but vary the amount of preheating by decreasing and increasing the acetylene pressure before each cut. Observe the finished cut for smoothness, melting of the top edge of the plate, amount of slag on the bottom edge of the plate, and ease of removal of this slag. Compare the plates cut and determine which amount of preheat produces the best results.
8. Make more cuts but vary the rate of travel from very slow to normal to very fast. Observe these finished cuts and check the appearance of the top and bottom of each plate, and also the ease of slag removal. Determine which rate of travel produces the best results.
9. Make more cuts but vary the amount of cutting oxygen pressure from low to normal to high and check the results as in step 8.
10. Make one or two cuts with the tip perpendicular to the work but move the torch so that the tip zigzags along the straight line drawn on the plate. Notice that the surface of the cut edge follows the amount and direction the tip moves from the straight line.

Fig. 10-3 Straight-line manual cut

Fig. 10-4 Straight-line machine cut

Basic Oxyacetylene Welding

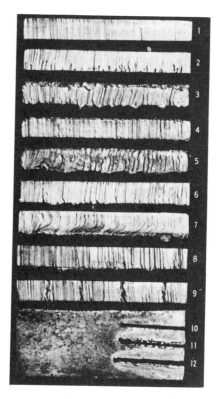

1. This is a correctly made cut in 1-in. plate. The edge is square and the draglines are vertical and not too pronounced.
2. Preheat flames were too small for this cut with the result that the cutting speed was too slow, causing gouging at the bottom.
3. Preheat flames were too long with the result that the top surface has melted over, the cut edge is rough, and there is an excessive amount of adhering slag.
4. Oxygen pressure was too low with the result that the top edge has melted over because of the too slow cutting speed.
5. Oxygen pressure was too high and the nozzle size too small with the result that the entire control of the cut has been lost.
6. Cutting speed was too slow with the result that the irregularities of the draglines are emphasized.
7. Cutting speed was too high with the result that there is a pronounced break to the dragline and the cut edge is rough.
8. Torch travel was unsteady with the result that the cut edge is wavy and rough.
9. Cut was lost and not carefully restarted with the result that bad gouges were caused at the restarting point.
10. Correct procedure was used in making this cut.
11. Too much preheat was used and the nozzle was held too close to the plate with the result that a bad melting over of the top edge occured.
12. Too little preheat was used and the flames were held too far from the plate with the result that the heat spread opened up the kerf at the top. The kerf is too wide at the top and tapers in.

Fig. 10-5 Common faults that occur in hand cutting

REVIEW QUESTIONS

1. A number of variables have been tried out in this unit. What conclusions can be drawn about the importance of each of these variables with regard to the ability to make straight line cuts?

    a. Tip angle

    b. Amount of preheat

    c. Amount of cutting oxygen pressure

    d. Rate of travel

    e. Direction of travel

Unit 11

BEVEL CUTTING

Making bevel cuts on steel plate is a common cutting operation. The technique is similar to that used for making straight cuts.

Skill in cutting operations is gained only through practice and with a definite goal in mind. Unguided wanderings over a plate add very little to an operator's skill and waste material and gas.

Materials

3/8-inch thick steel plate
Cutting tip, see chart 9-1 for size.

Procedure

1. Draw a series of parallel straight lines spaced on 2-inch centers on the work with soapstone.

    Note: These lines are sometimes centerpunched at close intervals to improve visibility. This improves accuracy.

2. Hold the cutting tip at an angle of 45 degrees with the work and keep this angle when bevel cutting.

3. Proceed with the cut in the same manner as in unit 10. The cut progresses better if the preheating orifices are aligned as in figure 11-1.

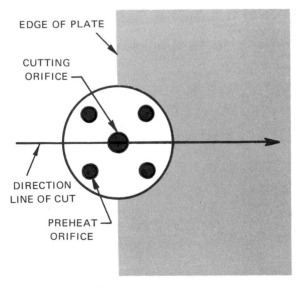

Fig. 11-1 Tip alignment for bevel cuts

Fig. 11-2 Manual bevel cut

Basic Oxyacetylene Welding

4. Inspect the finished cut for smoothness and uniformity of angle. Check the amount and ease of removal of slag.
5. Make more bevel cuts, but correct the variables until good results are obtained each time.
6. Make another cut, but as the tip moves forward, bring it alternately closer and farther from the work and observe the results.

REVIEW QUESTIONS

1. What variable is responsible for the slag being hard to remove from the bottom of the plate?

2. How does the distance of the tip from the work during the burning process affect the appearance of the finished cut?

3. In making the cut on the plate in this unit, how is the proper tip size, gas pressure, and rate of travel determined?

4. Why is a line to be cut center punched?

5. Why is bevel cutting done?

Unit 12

PIERCING AND HOLE CUTTING

Holes are easily cut in steel plate by the oxyacetylene flame-cutting procedure. This is a fast operation, adaptable to plates of varying thicknesses and holes of varying sizes. It is useful in cutting irregular shapes.

It is recommended that a diamond point chisel be used to turn up a burr at the point at which the cut is to be started. This burr reaches the kindling temperature much faster than the surface of a flat plate. If a large number of holes are to be pierced, this procedure saves large amounts of preheating gas and operator time.

CAUTION: If particular care is not used in this operation, molten metal may be blown in the face of the operator or into the tip of the torch.

Materials

3/8-inch thick steel plate
Cutting tip, see chart 9-1 for size.

Procedure

1. Pierce the plate, using the sequence of operations shown in figures 12-1 through 12-4.

   a. Hold the tip about 1/4 inch from the work until the surface reaches the kindling temperature, figures 12-1A and 12-2.
   b. Open the cutting oxygen valve slowly and, as the burning starts, back the tip away from the work to a distance of about 5/8 inch. The tip must be tilted slightly so that the oxide blows away from the operator and does not blow directly back at the tip, figures 12-1B, 12-1C, and 12-3.
   c. Hold the tip in this position until a small hole is pierced through the plate, figure 12-1D.
   d. Lower the tip to the normal burning distance and be sure that it is exactly square with the plate. Then move the torch to enlarge the hole, figures 12-1E and 12-4.

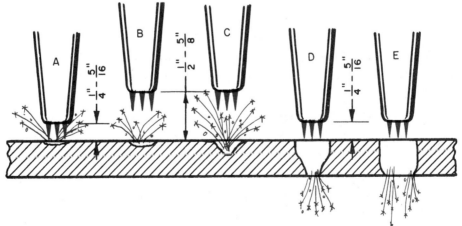

Fig. 12-1 Sequence for piercing plate

Basic Oxyacetylene Welding

Fig. 12-2 Preheating the plate

Fig. 12-3 Starting the cut

Fig. 12-4 Enlarging the hole

2. Continue to move the tip until the hole is the desired size.

   Note: This piercing procedure is recommended for two reasons:
   1. The tip is not ruined by burning the end. This occurs when the tip is held too close to the work for a long period.
   2. The possibility of the molten slag blowing back and clogging the cutting and preheating orifices is reduced.

3. After piercing the plate, move the tip in a circular path to cut a hole of the desired diameter. Considerable practice is necessary to become skillful in making holes which have straight sides, are reasonably round, and close enough to the given diameter to be acceptable.

4. With a pair of dividers, lay out some holes from 1/2 inch to 1 inch in diameter. Center punch the layout line at close intervals to serve as a guide for the cutting operation.

5. Cut the holes, but remember that when cutting to a line the plate should be pierced some distance inside the line. The hole can then be enlarged to the line and the operation completed.

6. Lay out and cut some holes 2 inches and 3 inches in diameter.

   Note: When making large diameter holes, pierce the plate and then move the torch in a straight line until the cut reaches the layout line. Then proceed with the circular cut.

7. Make some round discs. In this case, the piercing operation is performed at a distance outside the layout line. If these discs are to be turned to a specified diameter after cutting, enough material for this operation must be provided.

Piercing and Hole Cutting

REVIEW QUESTIONS

1. If a 3-inch round shaft is to be cut off, what preparation is necessary to insure a quick start of the cutting action?

2. How does the procedure vary from the above if the shaft to be cut is square instead of round?

3. If it is desirable to save both the disc and hole when cutting large diameter holes, what procedure should be followed?

4. Why is the tip tilted slightly when starting the cut?

5. Can hole piercing be dangerous for the operator?

Unit 13

WELDING SYMBOLS

DESCRIBING WELDS ON DRAWINGS

Welding symbols form a shorthand for the draftsman, fabricators, and welding operators. A few good symbols give more information than several paragraphs.

The American Welding Society has prepared a pamphlet, "Symbols for Welding and Nondestructive Testing" (AWS A2.4-76), which indicates to the draftsman the exact procedures and standards to be followed so the fabricators and welding operators may read and understand all the information necessary to produce the correct weld.

The standard AWS symbols for arc and gas welding are shown in figure 13-1.

EXAMPLES OF THE USE OF SYMBOLS

Each of the symbols in this unit should be studied and compared with the drawing which shows its significance. They should also be compared with the symbols shown in figure 13-1.

Throughout this book a symbol related to the particular job is shown together with its meaning. A study of each of these examples will clarify the meaning of the welding symbols.

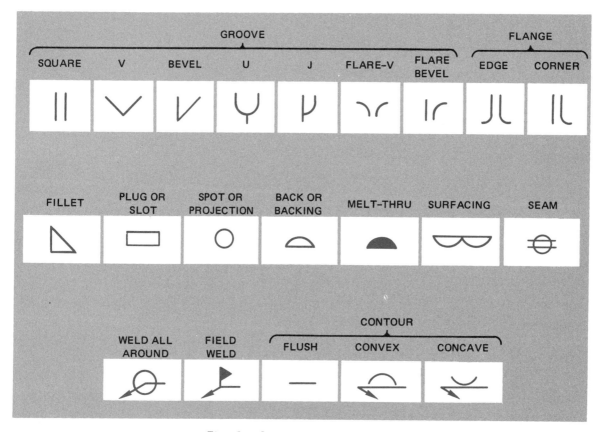

Fig. 13-1 Standard welding symbols

Welding Symbols

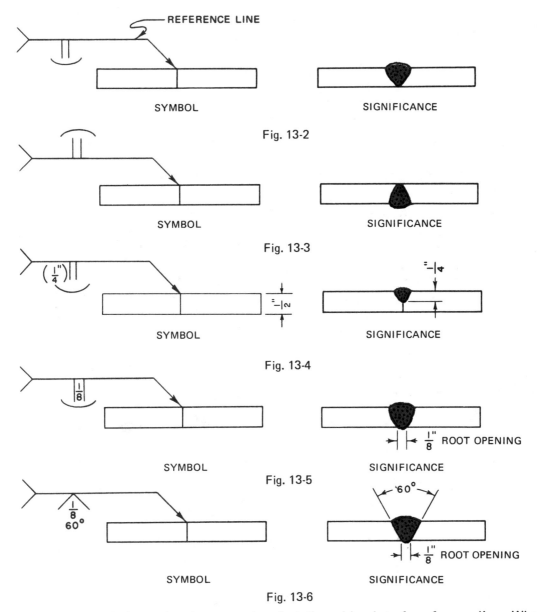

Fig. 13-2

Fig. 13-3

Fig. 13-4

Fig. 13-5

Fig. 13-6

The symbols from the chart are placed at the mid-point of a reference line. When the symbol is on the near side of the reference line the weld should be made on the arrow side of the joint as in figure 13-2.

If the symbol is on the other side of the reference line, as in figure 13-3, the weld should be made on the far side of the joint or the side opposite the arrowhead.

All penetration and fusion is to be complete unless otherwise indicated by a dimension positioned as shown by the (1/4) in figure 13-4.

To distinguish between root opening and depth of penetration, the amount of root opening for an open square butt joint is indicated by placing the dimension within the symbol, figure 13-5, instead of within parentheses as in the preceding drawing.

The included angle of beveled joints and the root opening is indicated in figure 13-6. If no root opening is indicated on the symbol, it is assumed that the plates are butted tight, unless the manufacturer has set up a standard for all butt joints.

37

Basic Oxyacetylene Welding

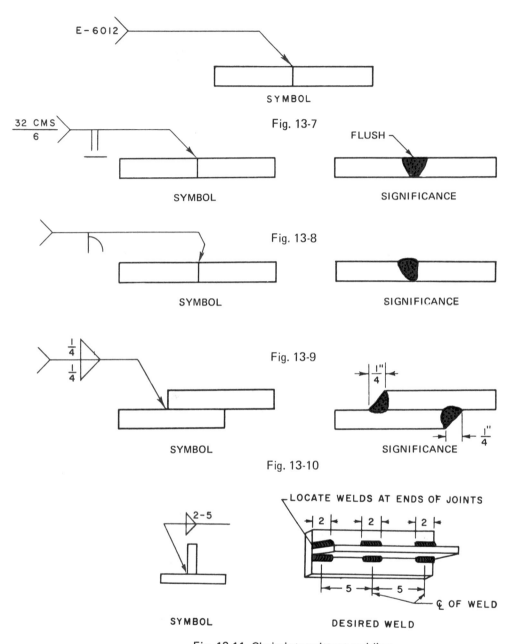

Fig. 13-7

Fig. 13-8

Fig. 13-9

Fig. 13-10

Fig. 13-11 Chain intermittent welding

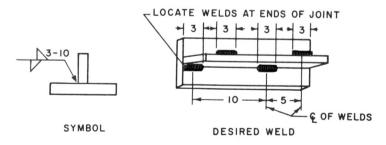

Fig. 13-12 Staggered intermittent welding

Welding Symbols

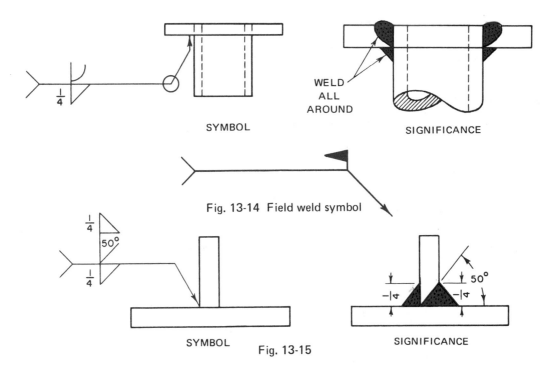

Fig. 13-14 Field weld symbol

Fig. 13-15

The tail of the arrow on reference lines is often provided so that a draftsman may indicate a particular specification not otherwise shown by the symbol. Such specifications are usually prepared by individual manufacturers in booklet or loose-leaf form for their engineering and fabricating departments. These specifications cover such items as the welding process to be used (i.e. arc or gas), the size and type of rod or electrode, and the preparation for welding, such as preheating.

Many manufacturers are using the AWS publication, "Symbols for Welding and Nondestructive Testing" which gives very complete rules and examples for welding symbols as well as a complete set of specifications with letters and numbers to indicate the process.

One method of indicating the type of rod to be used is shown in figure 13-7. This shows that the butt weld is to be made with an AWS classification E-6012 electrode.

In figure 13-8, the rod to be used is indicated as a number 32 CMS (carbon mild steel) type; the 6 indicates the size of the rod in 32nds of an inch. In this case it is a 3/16-inch diameter rod. In addition, the symbol indicates that the finished weld is to be flat or flush with the surface of the base metal. This may be accomplished by: G = Grinding, C = Chipping, or M = Milling.

When only one member of a joint is to be beveled, the arrow makes a definite break back toward the member to be beveled, figure 13-9.

The size of fillet and lap beads is indicated in figure 13-10. In all lap and fillet welds, the two legs of the weld are equal unless otherwise specified.

If the welds are to be chain intermittent, the length of the welds and the center-to-center spacing is indicated, as in figure 13-11.

When the weld is to be staggered, the symbol and desired weld is made as in figure 13-12.

An indication that the joint is to be welded all around is shown by using the weld all around symbol at the break in the reference line, as in figure 13-13.

Basic Oxyacetylene Welding

Field welds (any weld not made in the shop) are indicated by placing the field weld symbol at the break in the reference line, as in figure 13-14.

Several symbols may be used together when necessary, figure 13-15.

REVIEW QUESTIONS

1. What is the symbol for a 60-degree closed butt weld on pipe?

2. What is the symbol for a U-groove weld with a 3/32-inch root opening?

3. What is the symbol for a double V, closed butt joint in plate?

4. What is the symbol for a 1/2-inch fillet weld in which a column base is welded to an H-beam all around?

5. What is the symbol for a J-groove weld on the opposite side of a plate joint?

Unit 14

RUNNING BEADS AND OBSERVING EFFECTS

The quality of the finished weld depends to a large extent on the correct adjustment and use of the flame. This unit provides an opportunity to weld with different kinds of flames and to compare the results. At the same time some acutal welding skill is acquired.

Materials

16- or 14-gage mild steel, 2 to 4 in. wide X 6 to 9 in. long
Welding tip, see chart 6-2 for size.

Procedure

1. Light the torch and adjust the flame to neutral.

    Note: The right-handed operator will start on the right side of the workpiece and travel left (forehand technique). The left-handed operator will travel right.

2. Hold the tip of the inner cone of the flame about 1/8 inch above the work and pointed in the exact direction in which the weld is to proceed. The centerline of the flame should make an angle of 45 to 60 degrees with the work, figure 14-2.

3. Hold the flame in one spot until a puddle of metal 1/4 inch to 3/8 inch in diameter is formed.

4. Proceed with the weld, advancing the flame at a uniform speed in order to keep the molten puddle the same diameter at all times. This keeps the weld or *bead* the same width throughout its length. Start this bead 1/2 inch from the near edge of the plate being welded and proceed in a straight line parallel to this edge.

    Note: The width of the bead is directly related to the thickness of the plate being welded. The accepted standard for welds in aircraft tubing and light sheet metal requires the weld to be six times as wide as the thickness of the metal.

5. After the weld has been completed, examine it for uniformity of width and smoothness of appearance. Turn the plate over and examine the bottom for uniformity of *penetration*.

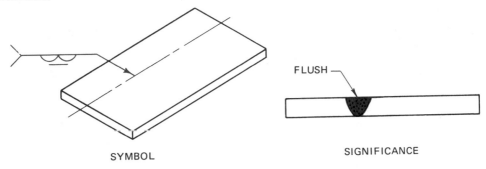

Fig. 14-1 Bead weld

Basic Oxyacetylene Welding

Fig. 14-2 Running a bead (Reprinted from Jeffus & Johnson, *Welding: Principles & Applications*, Figure 5-8)

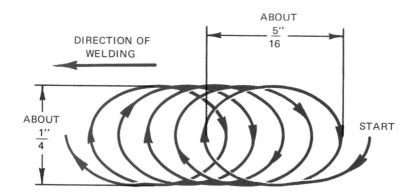

Fig. 14-3 Torch manipulation by a right-handed operator

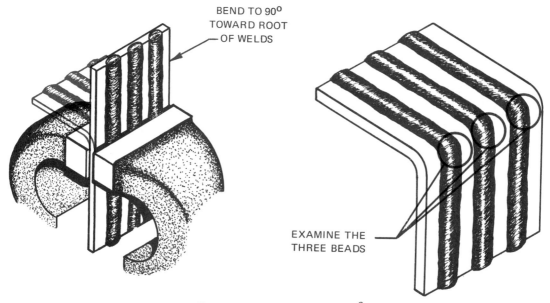

Fig. 14-4 Bending test weld 90°

Running Beads and Observing Effects

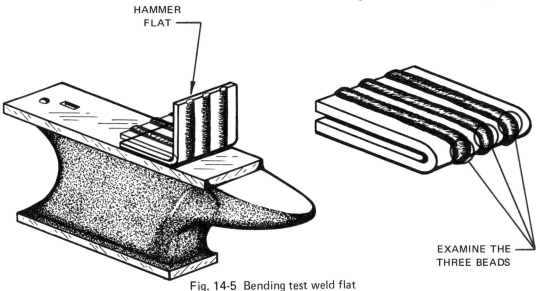

Fig. 14-5 Bending test weld flat

Note: The term penetration refers to the depth to which the parent metal is melted and fused. Fusion is the important factor. It is possible to obtain complete penetration of the base metal and less than complete fusion. The welder cannot expect to produce a welded joint equal in strength to the base metal if the fusion is less than 100 percent.

6. Make more beads on the same plate parallel to the first bead and 1/2 inch apart. Vary the angle the flame makes with the work for each new bead. Observe the finished welds for appearance and penetration.

7. Continue to make more beads on additional plates until good appearance and penetration are attained. Manipulate the flame to obtain better results. The simplest manipulation is to rotate the flame in a small circle with a clockwise motion so that the flame is alternately closer and farther away from the work, figure 14-3. The frequency and length of these cycles give the welder added control of the amount of heat applied to the work. In most cases, a better appearing weld results when the flame is manipulated in this way.

8. Run a bead on another plate, using a neutral flame. Then adjust the flame to carburizing and run a bead parallel to the first and about 1/2 inch from it. Note that the sparks coming from the molten puddle tend to break into bushy stars. Notice the cloudy appearance of the molten puddle.

   Note: When steel is heated in a carbon rich atmosphere, such as that produced by a carburizing flame, it tends to absorb the carbon in direct proportion to the temperature and the amount of carbon present. This carbon combines with the steel to form the hard, brittle substance known as iron carbide.

9. Adjust the flame to highly oxidizing and run a third bead parallel to the second and 1/2 inch from it. Note that the molten puddle is violently agitated and that the molten iron oxide has an incandescent frothy appearance. Iron oxide is a hard, brittle, low-strength material of no structural value. Note, also, that the oxidized bead is much narrower than either of the others.

Basic Oxyacetylene Welding

10. Cool the finished test plate and grasp it in a vise across the center of the three welds. Bend this plate 90 degrees toward the root of the welds, figure 10-4. Notice that the carburized bead has cracked and most of the oxidized material has fallen from the oxidized bead.

11. Hammer the test plate on an anvil until it is bent flat upon itself, figure 10-5. Notice that while the carburized and oxidized beads have cracked, the neutral bead has bent as much as the original material with no indication of failure.

REVIEW QUESTIONS

1. How does uniformity of procedure affect appearance of the finished weld?

2. What effect does flame angle have on penetration?

3. What effect does bead width have on fusion?

4. Does the force of the flame blow the molten puddle along the plate or does the molten puddle have to follow the direction the flame takes?

5. What type of flame is best for oxyacetylene welding operations? Why?

6. Why is the bead produced by the oxidizing flame narrower than that produced by the other two flames?

Unit 15

MAKING BEADS WITH WELDING ROD

In most oxyacetylene welding, additional metal is added to the weld by melting a filler rod into the puddle to produce a stronger weld. These rods are available in various diameters and materials.

The use of the filler rod requires the operator to manipulate not only the torch but also the rod. The proper coordination of the torch and the rod is necessary for the production of good welds. This unit provides an opportunity for practicing this manipulation and observing the results.

Materials

16- or-18-gage steel plate, approximately 4 in. X 9 in.
3/32-inch diameter steel welding rod
Welding tip, see chart 6-2 for size.

Procedure

1. Light the torch and adjust the flame to neutral.

2. Melt the base metal near one end of a plate until a puddle of the proper size is obtained as in unit 14.

3. Place the welding rod in the puddle, making sure the rod is aimed in the direction of travel of the weld, figure 15-2.

4. Proceed with the weld, making sure the welding rod and the tip of the torch make the correct angles with the work. Attempt to make a straight, uniform bead parallel to the edges of the base metal.

5. Make more welds in this manner but vary the angle that the rod makes with the work. Note the effect on the height of the bead.

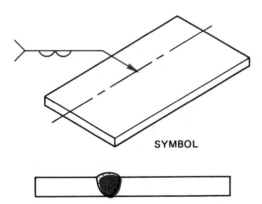

SYMBOL

SIGNIFICANCE

Fig. 15-1 Bead weld with welding rod

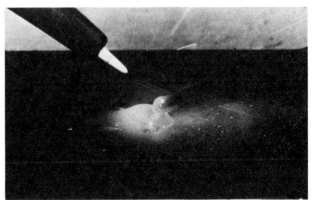

Fig. 15-2 Running a bead with filler rod (Adapted from Jeffus & Johnson, *Welding: Principles & Applications,* Figure 5-6)

Basic Oxyacetylene Welding

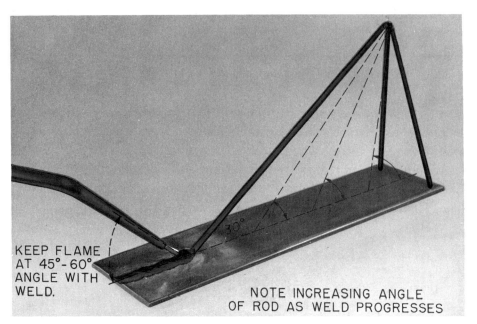

Fig. 15-3 Welding with fixed rod

6. Obtain another plate and set up the plate and rod according to figure 15-3.

   Note: The rod should be twice as long as the legs and welded to them. This produces a starting angle of 30 degrees between the rod and the welding line. As the weld progresses and the rod melts, filler metal is added at a faster rate. Observation of the finished weld shows the effect of *deposition rate* (rate at which filler metal is added) on the height of the finished bead.

7. Make the weld and observe that in this case, the rod manipulation is not necessary to make a weld with good appearance.

8. Obtain more plates and rods and make more welds; but, as the welding progresses, dip the rod end into the molten puddle with a regular rhythm, figure 15-4. Try one second in the puddle and one second out, and then increase the rhythm until the dipping action is rather rapid.

9. Observe the effect this dipping action has on bead height and uniformity.

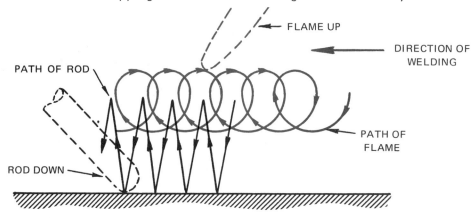

Fig. 15-4 Rod and flame manipulation by a right-handed operator

REVIEW QUESTIONS

1. What effect does rod angle have on the finished weld?

2. What effect does rod manipulation have on the weld?

3. What effect does uniformity of flame manipulation have on the weld?

4. What is wrong if the weld surface is flat in appearance?

5. What causes washout at the start and finish of a weld?

Unit 16

TACKING LIGHT STEEL PLATE AND MAKING BUTT WELDS

When making a butt weld, the metal expands as heat is applied and contracts as it cools. This may distort the metal and cause an unsatisfactory job.

To avoid such distortions, several precautions may be taken. One of the most common is to *tack* the two pieces in position. Tack welds are small temporary welds to hold the work in place and control the distortion. A skilled operator must know how to place tack welds, and what effect they have on the finished job.

Materials

Two pieces of 16- or 14-gage mild steel plate, 1 1/2 to 2 in. X 6 in. each
3/32-inch diameter steel welding rod
Welding tip, see chart 6-2 for size.

Procedure

1. Place the plates in position on the welding bench as indicated in figure 16-2.
2. Make tack welds approximately every two inches from right to left. The distance between tacks may be greater for thicker plates.

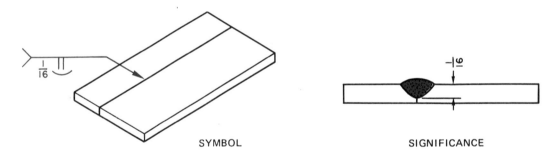

Fig. 16-1 Butt weld

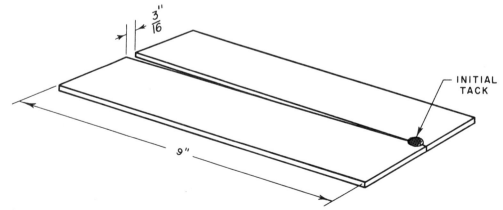

Fig. 16-2 Positioning plates for tacking

Tacking Light Steel Plate and Making Butt Welds

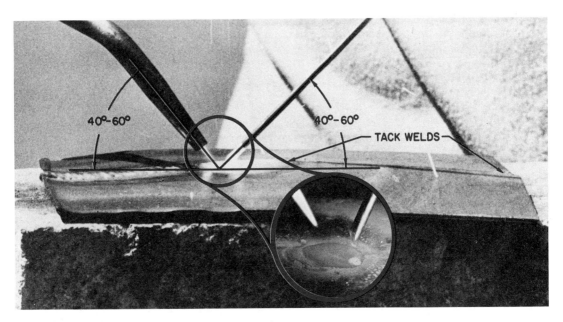

Fig. 16-3 Butt weld on 16 gage (Insert is adapted from Jeffus & Johnson, *Welding: Principles & Applications,* Figure 5-19)

3. Make the butt weld in much the same manner as in unit 15. However, the weld must be straight and the center of the bead must be on the exact center of the joint.
4. Examine the finished weld for uniformity; inspect the reverse side for penetration.
   Note: On light metal, penetration should be complete from one side of the plate. If this penetration is not obtained, secure more plates and make additional butt welds with wider beads. Practice this until penetration is complete. This happens when the width of the bead is about six times the thickness of the plate.
5. Test the butt welds, as shown in figure 16-4, by holding the finished weld in a vise with the centerline of the weld 1/8 inch above the jaws. Hammer the plate toward the face of the weld. A good weld shows no evidence of root cracks.
6. Obtain two more plates and tack both ends. Then try a third tack weld midway between the first two. Observe the effect of this procedure on plate alignment and ease of tacking.

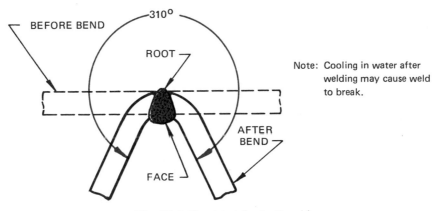

Fig. 16-4 Bend test for butt weld

Basic Oxyacetylene Welding

REVIEW QUESTIONS

1. What happens if the plates are placed in contact for their entire length and then tacked?

2. What effect does plate thickness have on plate spacing?

3. What effect does weld width have on penetration?

4. What effect does tacking only the ends of the joint have on plate alignment during welding?

5. What is penetration?

Unit 17

FLAT CORNER WELDS

The welded shape in this unit is easily tested by a simple method to determine the quality of the fusion. Welds may be tested to discover poor welds and errors in the procedure used.

Materials

Two pieces of 16- or 14-gage steel plate, 1 1/2 to 2 in. X 6 in. each
Welding tip, see chart 6-2 for size.

Procedure

1. Set up and tack the plates every two inches starting at the end.
2. Weld the plates by placing the flame on the work so that it is split by the sharp corner of the assembly, figure 17-4.
3. Make the weld, trying at all times to make a smooth uniform bead, figure 17-4.
4. Examine the finished bead for uniformity and complete penetration.
5. Check the finished bead by placing the assembly on an anvil and hammering the bead until the plates lie perfectly flat, figure 17-5. Examine the underside for cracks and lack of fusion.
6. Weld more joints of this type, varying the size of the puddle until complete penetration is obtained.

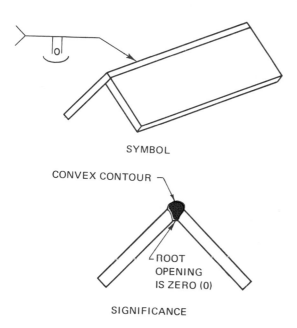

Fig. 17-1 Corner joint

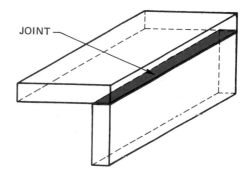

Fig. 17-2 Outside corner weld

Basic Oxyacetylene Welding

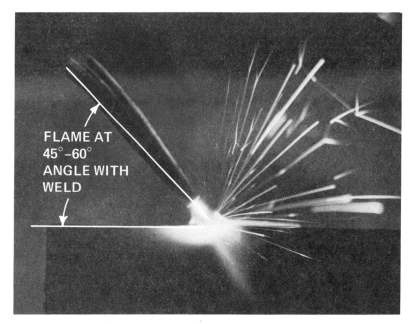

Fig. 17-3 Welder's eye view (Adapted from Jeffus & Johnson, *Welding: Principles & Applications,* Figure 5-26)

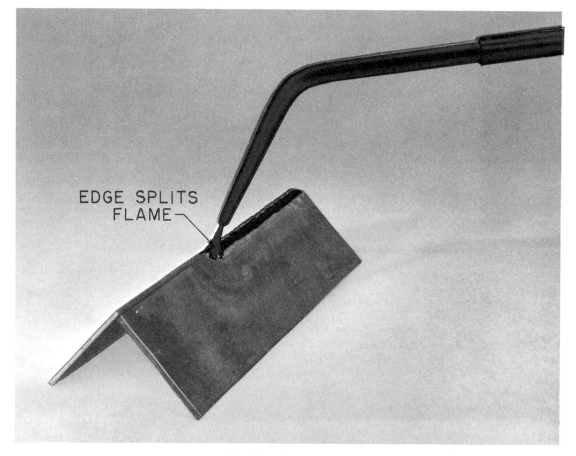

Fig. 17-4 Corner weld

Flat Corner Welds

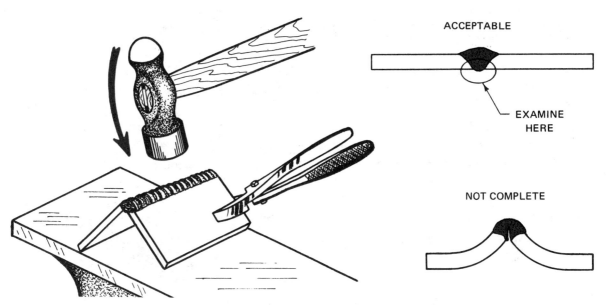

Fig. 17-5 Testing the weld

Note: When welding mild steel with a neutral flame, one way to check for complete penetration is the number of sparks coming from the molten puddle. Very few sparks are produced when the penetration is not complete, and large numbers of sparks are noted when the penetration is excessive. Good welders learn to use this indication to insure proper welds.

REVIEW QUESTIONS

1. In hammer-testing this type of joint, what color is the break if the penetration is not complete?

2. How does the strength of a properly made outside corner weld compare with the base metal?

3. What effect does too much penetration have on the appearance of the finished bead?

4. Is the angle of the torch tip critical, when making this weld? Why?

5. What is the function of a tack weld?

Unit 18

LAP WELDS ON LIGHT STEEL PLATE

The lap weld introduces some new difficulties from possible uneven melting of the two lapped plates. This type of welded joint illustrates the importance of the proper distribution of heat on the surface to be welded. One part of the joint may be melting while the other part may be far below melting temperature. This uneven heating prevents good fusion.

A simple test of the welded lap joint shows the quality of the weld. This shows the student where more practice is needed, and which welding procedures need to be changed to produce a better weld.

Materials

Two pieces of 16-gage mild steel plate,
  2 in. X 6 or 9 in. each
3/32-inch diameter steel welding rod
Welding tip, see chart 6-2 for size.

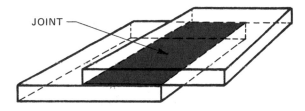

Fig. 18-1 Lap joint

Procedure

1. Place the plates on the welding bench in the position shown in figure 18-3.

   Note: Tack weld the two pieces at the ends so the tacks will not interfere with the weld.

2. Weld the plates, holding the rod and flame as shown in figure 18-3.

   Note: The edge of a plate melts more readily than the center of the plate. Therefore, in this weld there is a tendency for the top plate to melt back too far. This is overcome by placing the rod in the puddle as shown in figure 18-4. The rod is tilted slightly toward the top plate. The rod then absorbs some of the heat and eliminates excessive melting of the top plate.

3. Examine the finished weld for uniformity and for excessive melting of the top plate.

4. Place the weld in a vise, figure 18-5, and test by hammering the lapping plate until it forms a T with the bottom plate.

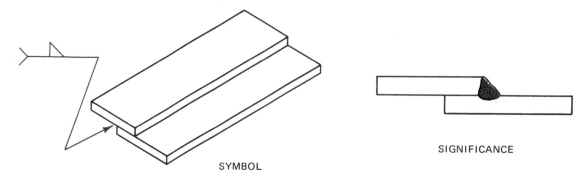

Fig. 18-2 Lap weld

Lap Welds on Light Steel Plate

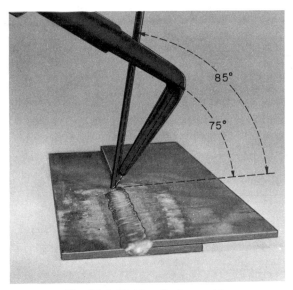

Fig. 18-3 Making the lap weld

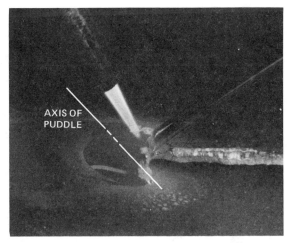

Fig. 18-4 Welder's eye view (Adapted from Jeffus & Johnson, *Welding: Principles & Applications,* Figure 5-20A)

Fig. 18-5 Testing the weld

5. Examine the root of the weld for complete penetration.
6. Practice this type of joint until good surface appearance is obtained and root penetration is complete.

REVIEW QUESTIONS

1. Where should the greatest amount of heat be directed in the lap weld?

2. What effect does the position of the rod in the puddle have on the melting of the lapping plate?

3. What is the relative position of the top and bottom of the molten puddle while the weld is being made?

4. Are the rod and torch angles more or less critical in this job than in the previous jobs?

Unit 19

TEE OR FILLET WELDS ON LIGHT STEEL PLATE

A tee or fillet weld on light steel plate provides experience in welding two steel plates set at right angles to each other. The angle of the flame to each of the two plates is important. The positions of the puddle and the rod also have an important effect on the quality of the weld. This job presents a new problem for the beginner — the possibility of *undercutting*.

Materials

Two pieces of 16- or 14-gage steel plate, 1 1/2 or 2 in. X 9 in. each
3/32-inch diameter steel welding rod
Welding tip, see chart 6-2 for size.

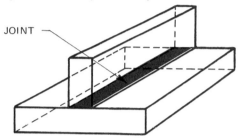

Fig. 19-1 Tee joint

Procedure

1. Set up and tack the plates as shown in figure 19-4.
2. Establish the size and shape of the weld.
3. Proceed with the welding, and pay particular attention to the following points:

    a. The centerline of the flame should make an angle of 45 degrees or less across the bottom plate.

    b. The angle the flame should make with the weld centerline varies from 60 degrees to 80 degrees. For thicker plate the flame should be pointed more directly into the weld.

    c. The puddle of molten metal should be positioned so that the bottom of the puddle is slightly ahead of the top. This is done by rotating the flame in a clockwise direction so the flame follows an oval path.

    d. The rod is usually placed near the top of the puddle so that it comes between the flame and the upstanding plate. In this position, the rod absorbs some of the heat and prevents excessive melting (burning through), or undercutting the vertical leg.

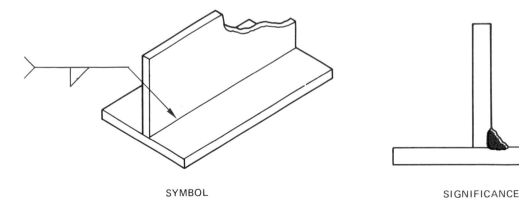

SYMBOL

SIGNIFICANCE

Fig. 19-2 Tee weld

Tee or Fillet Welds on Light Steel Plate

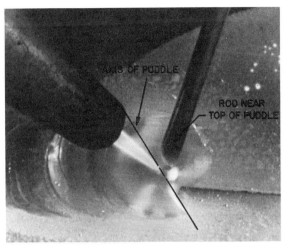

Fig. 19-3 Welder's eye view

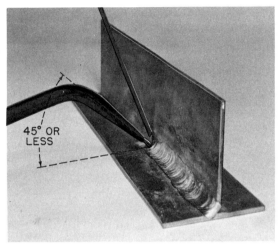

Fig. 19-4 Making the fillet weld

Note: Undercutting may be encountered, figure 19-5. This is an absence of metal along the top edge of the weld. It is caused by too much heat or poor rod movement. It should be avoided at all times.

4. Check the finished fillet weld by placing the assembly on an anvil and hammering the upstanding leg flat toward the face of the weld, figure 19-6. Examine for cracks in the root of the weld.

5. Make more joints of this type until smooth uniform welds are made. It should be possible to bend these welds 90 degrees in either direction without cracking.

6. After a fillet weld has been made on one side of the assembly, make a weld on the opposite side.

   Note: The first weld has produced oxide or scale on the reverse side. This is removed easily by playing the flame rapidly back and forth along the back surface of the joint. The flame causes the oxide to expand and pop from the surface.

CAUTION: When using the flame descaling procedure, extreme care must be used to protect the eyes and skin from burns caused by the hot, flying scale.

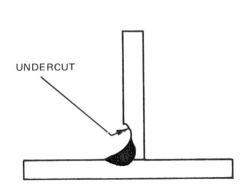

Fig. 19-5 Undercutting

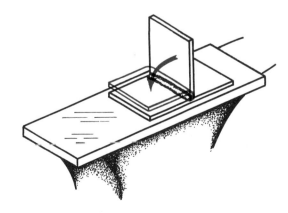

Fig. 19-6 Testing the fillet weld

Basic Oxyacetylene Welding

REVIEW QUESTIONS

1. Is it possible to develop the full strength of the joint when undercutting is present? Explain.

2. What effect does flame and rod movement have on root penetration and appearance of the weld?

3. When making fillet welds on both sides of the joint, how does the amount of heat required for the second weld compare to the first? Why?

4. What effect does melting the excess oxide on the plates and fusing it with the second weld (in step 6) have on the finished bead?

5. Define undercutting.

Unit 20

BEADS OR WELDS ON HEAVY STEEL PLATE

Although the principles of welding on heavy steel plate are the same as with lighter plate, the problems are greater because more heat is required. To distribute this heat properly, attention must be paid to torch motion and flame angle.

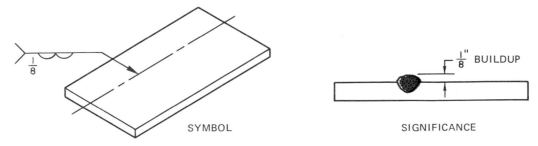

Fig. 20-1 Bead on heavy steel plate

Materials

    3/16-inch thick mild steel plate, 4 in. X 9 in.
    1/8-inch diameter steel welding rod
    Welding tip, see chart 6-2 for size.

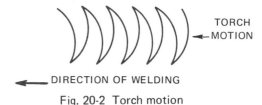

Fig. 20-2 Torch motion

Procedure

1. Adjust the flame to neutral.
2. Apply the flame to the work with the tip at an angle of 75 to 80 degrees along the line of the weld.
3. Weld a bead the length of the plate as in unit 15. The flame should be moved in a half-moon weave to produce a weld of adequate width and depth of penetration, figure 20-2. Make the bead 1/2 inch to 5/8 inch wide.
4. Observe the finished bead for appearance, particularly the spacing of ripples, edge of fusion, and penetration, figure 20-3.

Fig. 20-3 Weave beads with rod

Basic Oxyacetylene Welding

5. Run more beads on the same plate by varying the angle that the tip makes with the work with each bead.
6. Observe these beads for uniformity. Determine the effect of too little or too great a tip angle on the penetration of the base metal and the appearance of the finished weld.

   Note: This job requires a much larger tip size than any used in previous jobs. As a result, the gas consumption rises very rapidly with a corresponding rise in the hourly cost of operation. The gas should be shut off as soon as the welding is completed so that the cost of operation can be kept down.

REVIEW QUESTIONS

1. What effect does flame angle have on the size and shape of the puddle and the bead ripples?

2. Draw a sketch of the puddle when the flame angle is too small.

3. What effect does flame angle have on penetration in heavy plate?

4. Is the tip manipulation the same for heavy plate as it is for light plate?

5. Does penetration become more of a problem on heavy plate? How is it helped?

Unit 21

MANIPULATION OF WELDING ROD ON HEAVY STEEL PLATE

Considerable practice is required to develop skill in manipulating the welding rod and flame in the molten puddle when welding heavy steel plate. The relationships of the rod, flame, and puddle are particularly important. This unit provides practice in manipulating all three.

Materials

3/16- or 1/4-inch thick mild steel plate, 4 in. X 9 in.
1/8-inch diameter steel welding rod
Welding tip, see chart 6-2 for size.

Fig. 21-1 Torch and rod motion

Procedure

1. Apply a neutral flame to the work as in unit 20.
2. The molten puddle should be 1/2 inch to 5/8 inch wide.
3. Weld as in unit 20 except that both the rod and the flame should be moved alternately. In other words, the rod and flame should be moved so that they are on opposite sides of the molten puddle at all times, figure 21-1.
4. Inspect the finished weld for appearance.
5. Make more parallel welds on the same plate, but vary the angle of the rod and the flame. Observe the effect that this variation has on the height of the weld, the depth of penetration, and the face of the weld.
6. Compare these beads or welds with those made in unit 20.

REVIEW QUESTIONS

1. What is the advantage of moving the flame and rod, rather than the flame alone?

2. Which weld has the more uniform appearance?

3. Can a weld bead on heavy plate be too large?

4. Will the surface appearance of the bead be more uneven on heavy plate than light plate welding?

5. On heavy plate is the penetration the same as, less than or more than that on light plate if the correct tip is used?

Unit 22

BUTT WELDS ON HEAVY STEEL PLATE

Butt welding heavy steel plate is a basic welding operation. The quality of the weld is determined by: the positioning of the plates, tacking the plates, preparation of the edges, and the fit of the plates to each other. All of these factors also affect the ease with which the weld may be made.

Besides providing another opportunity to acquire more skill in butt welding, this unit points out the importance of plate edge spacing.

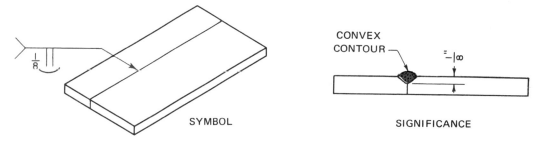

Fig. 22-1 Butt weld on heavy steel plate

Materials

Two pieces of 3/16-inch or 1/4-inch thick mild steel plate, 1 1/2 to 2 in. X 9 in. each
1/8-inch diameter steel welding rod
Welding tip, see chart 6-2 for size.

Procedure

1. Align the plates and tack as in making butt welds in thin plate, unit 16. The distance between tacks may be greater here than on the thinner material.

2. Proceed with welding as in unit 21.

3. Obtain more plates and tack them so that they are spaced 1/16 inch apart for one pair, 3/32 inch apart for another pair, and 1/8 inch apart for a third pair, figure 22-2.

   Note: When the plate edges are touching as in step 1, the joint is called a *closed square butt joint*. When they are spaced as in step 3, the joint is an *open square butt joint*.

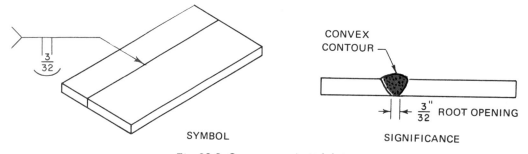

Fig. 22-2 Open square butt joint

Butt Welds on Heavy Steel Plate

4. Weld each pair of plates.
5. Examine and compare the finished butt joints.

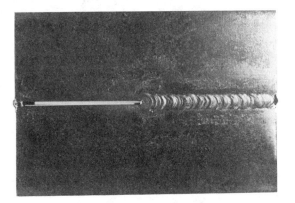

Fig. 22-3 Butt weld on heavy plate

REVIEW QUESTIONS

1. What is the relationship between plate thickness and distance between tacks?

2. What effect does plate edge spacing have on:
    a. Penetration?

    b. General appearance of the welds?

3. Draw a sketch of a cross section of a closed square butt joint.

4. Draw a sketch of a cross section of a open square butt joint.

5. Why is there a tendency for the open square butt joint to melt away at the plate edges?

63

Unit 23

LAP WELDS ON HEAVY STEEL PLATE

The procedure for making a lap or fillet weld on heavy steel plate is much the same as that required for making this weld on light plate.

In the job performed in this unit, more metal must be deposited because of the thicker plate. Since the heat must cover a larger area, more torch movement is involved. In addition, the larger molten puddle requires more rod movement.

Materials

Two pieces of 3/16- or 1/4-inch thick steel plate, 2 in. X 9 in. each
1/8-inch diameter steel welding rod
Welding tip, see chart 6-2 for size.

Procedure

1. Align the plates so that they lap approximately halfway in the long direction, figure 23-1.

    Note: Tack weld the two pieces at the ends as in unit 18.

2. Weld the plates, but point the flame toward the joint more than when lap-welding light steel plate, figures 23-2 and 23-3. Keep the flame pointed more toward the top plate. The tendency to overmelt the top plate is much less than in welding light plate.

3. As the welding proceeds, rotate the flame clockwise so that the bottom of the molten puddle is slightly ahead of the top. Try varying the position of the rod in the molten puddle.

4. Obtain more plates and repeat this type of joint; but alternate the flame and rod in the puddle in much the same manner as in the butt joint. Manipulate the flame and rod to keep the bottom of the puddle slightly in advance of the top.

5. Make more joints of this type. In each weld, vary the amount that the bottom of the puddle leads the top.

6. Inspect the finished welds for appearance.

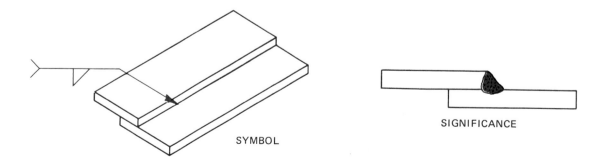

SYMBOL

SIGNIFICANCE

Lap Welds on Heavy Steel Plate

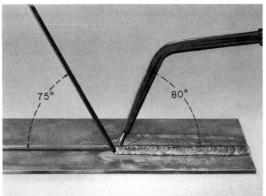

Fig. 23-2 Welder's eye view

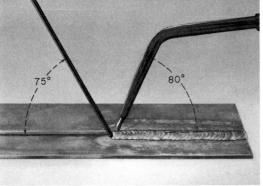

Fig. 23-3 Lap weld on heavy steel plate

7. Break these joints in a vise and examine the welds. Inspect for complete fusion of both plates. Note the size of the crystals or grain of the weld metal in the break.

8. In a piece of steel about the same size, make a saw cut part way through, so it breaks when bent. Compare the grain structure or crystal size of this break with the grain structure of the broken weld metal.

REVIEW QUESTIONS

1. What effect does the tip angle have on the appearance of the finished bead?

2. What effect does the tip angle have on the fusion of the two plates?

3. What effect does the heat of fusion have on the size of the grain structure in the weld and nearby metal?

4. Does undercut become more of a problem with heavy plate? Why?

5. Is it possible for a lap weld to look good and not be strong? Explain.

Unit 24

FILLET WELDS ON HEAVY STEEL PLATE

Fillet or tee joints are welded in much the same way as the lighter plates in unit 19. However, changes must be made in the flame angle, rod position, and the molten puddle to produce a good weld.

Materials

Two pieces of 3/16- or 1/4-inch thick mild steel plate, 2 in. X 9 in. each
1/8-inch diameter steel welding rod
Welding tip, see chart 6-2 for size.

Procedure

1. Position the plates so that the upstanding plate is about in the center of the flat plate. Tack the plates so that the upstanding leg makes an angle of exactly 90 degrees with the bottom plate.

2. Make the weld, using much the same technique as when making the weld in unit 23. The flame angle and rod position are very important and must be correct to avoid undercutting the upstanding leg, figure 24-2.

3. Check the angle between the vertical and horizontal plates after the weld has cooled.

4. Make more joints, tacking the plates at slightly varying angles. This will allow for shrinkage as the weld cools. Check the angle between the vertical and horizontal plates after the weld has cooled.

5. Make more joints of this type, varying the flame angle, rod position, and amount of lead of the bottom of the puddle.

   Note: The size of the legs of the weld must equal the thickness of the plate being welded if the joint is to be full strength.

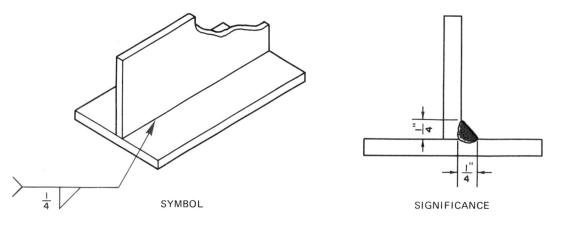

Fig. 24-1 Fillet weld

Fillet Welds on Heavy Steel Plate

Fig. 24-2 Fillet weld on heavy steel plate

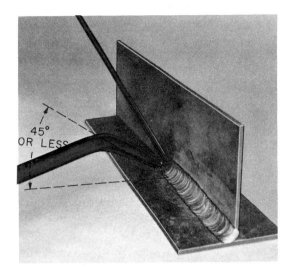

Fig. 24-3 Welder's eye view

6. Break the joints by placing the bottom plate in a vise and bending the upstanding leg toward the weld face.//
7. Examine the welds for fusion and penetration, especially at the root of the weld.

REVIEW QUESTIONS

1. What effect does flame angle have on the appearance of the finished bead?

2. What effect does flame angle have on the tendency to undercut?

3. What effect does rod position in the molten puddle have on appearance and the tendency to undercut?

4. How are the plates set up so the final angle is exactly 90 degrees, after shrinkage? Make a cross section sketch of this joint as set up ready for welding.

5. Why is correct hand protection so important when making this weld?

Unit 25

BEVELED BUTT WELD ON HEAVY STEEL PLATE

Complete fusion and penetration are of great importance to the welding operator. Joints which have been V'd or beveled make penetration in heavy plate much easier. This unit provides practice in flame cutting, flame and rod movement, and multilayer welding.

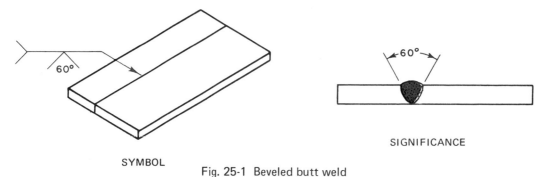

SYMBOL

SIGNIFICANCE

Fig. 25-1 Beveled butt weld

Materials

    Two pieces of 5/16- or 3/8-inch thick mild steel plate, 3 in. X 9 in. each
    1/8-inch diameter steel welding rod
    Welding tip, see chart 6-2 for size.

Procedure

1. Flame-cut one edge of each of the plates so that the included angle is 60 degrees.
2. Align and tack the plates with 3/32-inch root opening, figure 25-1.
3. Make the first weld in the bottom of the groove. Fusion between the plates should be complete. The flame should make an angle of 60 to 75 degrees with the work. If there is a tendency to burn through at the bottom of the V, try a smaller flame angle. If the bead is irregular and the fusion is poor, try a greater flame angle.

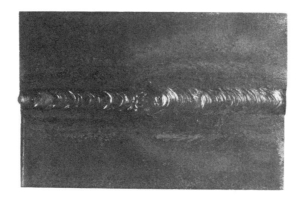

Fig. 25-2 Beveled butt weld

Beveled Butt Weld on Heavy Steel Plate

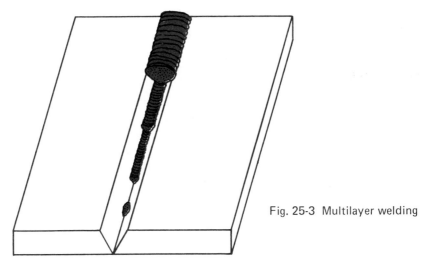

Fig. 25-3 Multilayer welding

4. Apply the second bead as shown in figure 25-3. Increase the flame angle and weave the torch slightly. Apply only enough rod to make a weld that is flat or somewhat concave.

5. Make the third bead, weaving the flame and rod alternately, as in step 3, unit 21. Control of the flame is important at this point so the fusion is complete along the top edges of the V. However, the melting must not be so great as to make the weld too wide. The weld should not be over 1/8 inch wider than the top of the opening.

6. Make two saw cuts across the finished weld to produce a sample 1 1/2 inch wide. Examine these sawed surfaces for complete fusion and absence of gas pockets or holes in the weld.

7. Break this sample by hammering it toward the face of the weld. Examine the break for lack of fusion, oxide in the weld, and gas pockets.

REVIEW QUESTIONS

1. What effect does too small a flame angle have on the first bead?

2. How can you overcome the tendency of the first and second beads to pile up or become convex?

3. How do the flame and rod motion, and the flame angle in this unit compare with those in unit 21?

4. How can poor fusion at the root of the weld be corrected?

5. Is the root spacing of the joint critical for good penetration? Why?

Unit 26

BACKHAND WELDING ON HEAVY STEEL PLATE

All of the welds made so far have been made by the *forehand* method with the flame pointing in the direction of welding. In the *backhand* method, the torch and rod are held in the same position but the flame points opposite the direction of welding. The welding flame is directed at the completed portion of the weld. The welding rod is placed between the completed weld and the flame.

Materials

Two pieces of 1/4-inch thick mild steel plate, 3 in. X 9 in. each
1/8-inch diameter steel welding rod
Welding tip, see chart 6-2 for size.

Procedure

1. Align and tack the plates as shown in figure 26-1.
2. Start a puddle at the end of the joint and proceed with the weld by the backhand method. Use a weaving motion similar to that in units 21 and 25.
3. As the weld progresses, observe the bead and alter the flame angle until the weld has good appearance.
4. Make joints of this type until welds are produced with uniform ripples and complete fusion throughout the entire length.

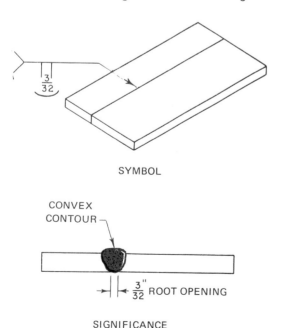

Fig. 26-1 Butt joint in heavy plate

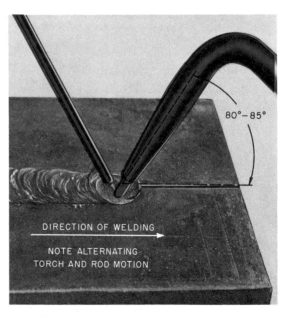

Fig. 26-2 Welder's eye view — backhand welding by a right-handed operator

5. As these welds proceed, observe the amount of penetration and the width of the beads. Compare these welds with those made in unit 21.

6. Make two saw cuts across the finished weld to produce a sample 1 1/2 inches wide. Examine these sawed surfaces for complete fusion and absence of gas pockets in the weld.

7. Break this sample by hammering it toward the face of the weld. Examine the break for lack of fusion, oxides in the weld, and gas pockets.

REVIEW QUESTIONS

1. How does the width of a backhand weld compare with the width of a forehand weld on a plate of similar thickness?

2. How does the flame angle compare with forehand welding?

3. Sketch a lengthwise cross section of this type of weld showing the finished bead, the rod, and flame penetration.

4. What is the advantage of backhand welding over forehand?

5. Which type of welding is faster, forehand or backhand?

Unit 27

BACKHAND WELDING OF BEVELED BUTT JOINTS

Backhand welding of beveled butt joints is used to great advantage in oxyacetylene welding of steel pipe. This unit provides practice in making this type of joint. Although this joint is not easy to make, it is much less difficult than a similar one in pipe. This joint should be mastered before pipe joints are welded.

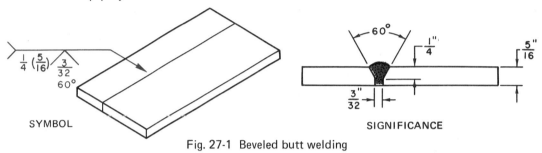

Fig. 27-1 Beveled butt welding

Materials

Two pieces of 5/16- or 3/8-inch thick mild steel plate, 3 in. X 9 in. each
3/16-inch diameter steel welding rod
Welding tip, see chart 6-2 for size.

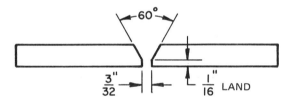

Fig. 27-2 Cross section of beveled plates

Procedure

1. Bevel the plates as indicated in figure 27-1.
2. If equipment is not available to produce the *land* shown in figure 27-2, bevel the plates to a feather edge and grind them to obtain the amount of land indicated. The land is the vertical part of the opening.
3. Align and tack the plates as shown.
4. Weld the root by the backhand method so that fusion is complete. This is done by using the proper flame angle and rod application. The flame angle must be steep enough to insure good penetration. The rod may have to be dipped in and out of the puddle to allow the heat to melt the root of the base metal.
5. Make the second pass, using the motion described in unit 21. Be sure the bead is completely fused, and that it is not over 1/8 inch wider than the top of the original groove, figures 27-3 and 27-4.
6. Cut a section from this weld and examine it for penetration and fusion.
7. Weld additional plates with one pass only on each bead.
8. Cut a section from each of these joints and inspect them as in step 6. Check especially for penetration, fusion, and bead appearance.

Backhand Welding of Beveled Butt Joints

Fig. 27-3 Welder's eye view

Fig. 27-4 Beveled butt welding — backhand method

REVIEW QUESTIONS

1. What should be the shape of the surface of the root pass when making the joint at step 4?

2. Why does step 5 caution that the face of the bead must be kept narrow?

3. What is the difficulty in making this joint with a single pass?

4. Does it take longer to make the joint by the forehand method or the backhand method?

5. Will the welder feel more heat or less heat in backhand welding?

Unit 28

BRAZING WITH BRONZE ROD

Brazing is a process in which metals are joined at a temperature greater than 800 degrees F. The base metal has a melting point at least 50 degrees F. higher than the filler rod. This indicates that:

- The base metal is not melted during this process.
- The joint is held together by the adhesion of the brazing alloy to the base metal rather than by cohesion. Cohesion takes place when the base metal and filler rod are fused.
- A brazed joint is bonded rather than welded. Brazing is used primarily for repairing cast iron parts and joining very light sheet metal. The primary alloys of brass are copper and zinc. The primary alloys of bronze are copper and tin.

This unit defines the techniques involved in oxyacetylene torch brazing using bronze filler rods. The action of flux and bronze during the brazing process will be examined.

Materials
16- to 11-gage thick clean steel plates, 2 in. X 6 in. each
1/8-inch diameter bronze rod or coated bronze rod
Welding tip one size larger than for welding a similar plate
Suitable dry-type brazing flux

Procedure
1. Place a piece of *clean* steel plate on the welding bench so that one end overhangs the bench.

   Note: The word clean refers to steel which has all the mill scale or iron oxide removed by either chemical or mechanical means. Mill scale and rust makes the production of strong joints either very difficult or impossible.

2. Sprinkle some flux on this plate. Apply the flame to the bottom side of the plate until the flux melts and flows over the plate.

3. Observe the color of the plate at this temperature. Also note that the flux flows freely. These two factors are the best guides to the proper brazing temperature.

4. Cool the plate and observe the metal under the flux. Compare the color of this metal to that of the unheated part of the plate. Note that the fluxed part of the plate is much whiter in color, indicating that the flux has cleaned the metal. The primary purpose of the flux is to chemically clean the surface for the brazing alloy. A secondary purpose is to protect the finished bead from the atmosphere during the brazing and cooling period.

5. Put some flux and a drop of brazing alloy on the end of an overhanging plate and heat as before. Observe that the flux melts first, then the alloy.

6. Move the flame about on the bottom of the plate. Note that the alloy flows freely in all directions as long as the flux is flowing ahead and cleaning the metal.

Brazing with Bronze Rod

Note: The process of adhering a thin coating of bronze or some other metal to the surface of the base metal is called *tinning.*

7. Continue to apply heat. Note that the alloy starts to burn with a greenish flame, first in small spots and then in wider areas. At the same time, the alloy gives off a white smoke and leaves a white residue on the plate.

   Note: This residue is the zinc being overheated to the point where it evaporates and burns, to form zinc oxide. This heating is harmful to the brazed joint because the alloy is changed by the removal of the zinc which is replaced to some extent by the zinc oxide.

CAUTION: Breathing of zinc oxide may cause the operator to become violently ill.

8. Place a drop of the brazing alloy on the end of another overhanging plate and heat as before but do not use flux. Note that the alloy does not spread over the plate as before. Instead, it tends to vaporize and burn when the melting point is reached. This indicates that a brazed joint cannot be made without a dry-type brazing flux or one of the paste-type fluxes.

REVIEW QUESTIONS

1. What is the major difference between a brazed joint and a welded joint?

2. It is possible to braze brass or bronze if the proper alloy is used. What two conditions determine whether the joint is brazed or welded?

3. How does the flux act as a guide to the temperature of the joint?

4. From observations of the flowing characteristics of this alloy, what precaution should be taken to insure that the bead does not become too wide?

5. Can brazing create a toxic atmosphere for the welder? Explain.

Unit 29

RUNNING BEADS WITH BRONZE ROD

This job provides practice in flame movement so that the operator develops skill in making good beads of a given size and shape with bronze rod.

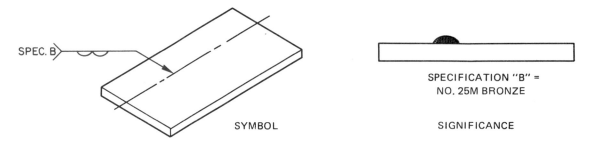

SYMBOL

SPECIFICATION "B" = NO. 25M BRONZE

SIGNIFICANCE

Fig. 29-1 Bead with bronze rod

Materials

16- or 14-gage steel plate, 2 in. X 6 in.
1/8-inch diameter bronze rod or coated bronze rod
Suitable brazing flux
Welding tip one size larger than for welding on similar plate

Procedure

1. Adjust the flame according to instructions. The flame for brazing varies with the alloy being used from slightly carburizing through neutral to slightly oxidizing.

2. Heat the rod with the flame until the flux clings to the rod when it is dipped in the flux container.

3. Apply the flame to both the rod and the work until a drop of alloy is left on the work. Remove the rod from the flame and continue to heat the plate until the drop melts and flows over an area of about 1/2-inch diameter.

4. Start a bead lengthwise on the plate. Keep the rod close to the flame and move both the rod and the flame in a spiral. Both the rod and flame are alternately close to the work and far away.

    Note: Check the flame angle. It is usually less than that used to weld a plate of similar size.

5. Bring the rod and flame in contact with the work when they are on the downswing of the spiral motion, figure 29-2.

6. Drag the rod in the direction of the brazing before removing it for the upswing. This draws the flux ahead of the molten alloy and speeds the cleaning process.

7. Continue with the brazing and note that the flux flows ahead of the alloy. When this no longer happens, dip the still hot rod in the flux and continue with the bead.

Running Beads with Bronze Rod

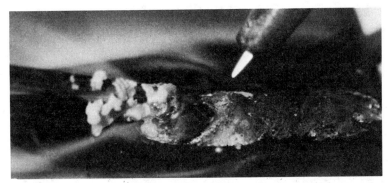
Fig. 29-2 The filler metal and the flame contact the work at the same time.

8. Inspect the finished bead for width, height, ripples, and for the white residue that indicates overheating.

9. Make more beads, using these variations in procedure:
   a. Change the size of circle made by the flame and rod.
   b. Apply the flame to the work for longer and shorter intervals of time.
   c. Increase and decrease the amount of flux on the rod.

10. Inspect the finished beads.

REVIEW QUESTIONS

1. What effect does too much or too little heat have on the appearance of the finished bead?

2. What effect does too much or too little flux have on the ease of brazing?

3. Is temperature control in the base metal more critical in brazing than in welding? Explain.

4. Is it possible to hold the inner cone of the flame farther away from the joint than when welding in order to make a narrower braze? Explain.

5. What is the color of the plate when it is at the proper temperature for brazing?

Unit 30

SQUARE BUTT BRAZING ON LIGHT STEEL PLATE

This job gives practice in making butt brazes on light steel plate. Although fusion welding and brazing are two different processes, they have many points in common. Thus, it is possible to use much of the experience gained in welding to produce braze joints.

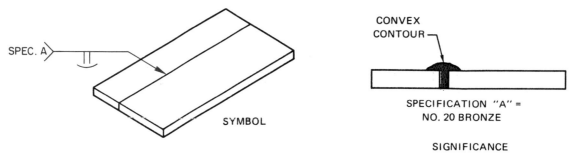

Fig. 30-1 Square butt braze

Materials

Two pieces 16- to 11-gage thick steel plate, 1 1/2 to 2 in. X 6 in. each
1/8-inch diameter bronze rod or coated bronze rod
Suitable dry brazing flux
Welding tip one size larger than for welding comparable steel plate

Procedure

1. Align the plates and tack with bronze. Be sure the edges of the plates are in contact along the entire length of the joint.
2. Adjust the flame for the alloy being used.
3. Braze as in unit 25. Try to keep the bead narrow.
4. Hold the plates in a vise and bend them until they break, following the procedure for welded joints.
5. Examine the broken joint for evenness and depth of bond.
6. Tack more plates, spacing the edges slightly farther apart. Braze as before.
7. Break these plates and examine the results.

REVIEW QUESTIONS

1. What effect does plate edge spacing have on the depth of bond?

Square Butt Brazing on Light Steel Plate

2. What is the procedure for getting a greater bond depth?

3. Is any tinning apparent on the reverse side of the joint brazed in this unit? If not, what conditions are necessary to obtain such tinning?

4. Is the square butt joint a good type of brazing joint? Why?

5. Is brazing stronger than fusion welding?

Unit 31

BRAZED LAP JOINTS

This job develops manipulative skill in making brazed lap joints on steel plate. Some of the procedures, problems and difficulties encountered are the same as those in welding lap joints on light steel plate.

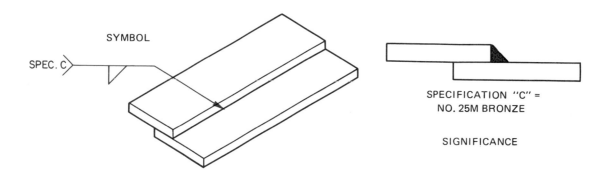

Fig. 31-1 Brazed lap joint

Materials

Two pieces of 16- to 11-gage thick clean steel plate, 2 in. X 6 in. each
1/8-inch diameter bronze rod
Suitable flux
Welding tip one size larger than for welding comparable plate

Procedure

1. Lap the plates following the procedure outlined for welding a lap joint, unit 18.
2. Adjust the flame for the alloy being used.
3. Proceed with brazing and observe the tendency of the alloy to flow on the top plate.
4. Change the angle the flame makes with the line of brazing to correct the tendency to over braze the top plate, figure 31-2.
5. Continue to make these joints until enough skill is acquired to make brazed lap joints with a good appearance.

    Note: Try to make all brazed beads about the same width as a weld made on material of similar thickness. The tinning action of the alloy causes the braze to become too wide if flame movement is not carefully controlled.

6. Break and examine these joints using the same procedure as that used to break welded lap joints.

Brazed Lap Joints

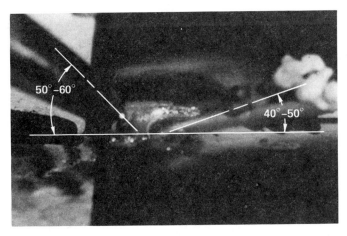

Fig. 31-2 Welder's eye view (Adapted from Jeffus & Johnson, *Welding: Principles & Applications,* Figure 7-27)

Fig. 31-3 Brazed lap joint

REVIEW QUESTIONS

1. How critical is the flame angle as compared to that used when welding a similar joint?

2. What effect does holding the inner cone too far from the work have on the width of the finished bead?

3. When brazing a lap joint, is there a tendency for the alloy to flow between the plates?

4. How does the grain size of the break in the brazed joint compare with that of a welded lap joint?

5. What is the strongest type of brazed joint?

Unit 32

BRAZED TEE JOINTS

This job helps the student obtain skill in the technique of making strong joints of good appearance using bronze filler metal.

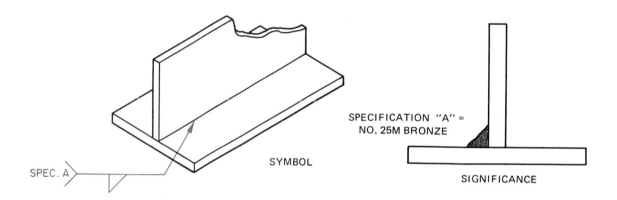

Fig. 32-1 Brazed tee or fillet joint

Materials

Two pieces of 16- to 11-gage thick clean steel plate, 2 in. X 6 in. each
1/8-inch diameter bronze rod
Flux
Welding tip one size larger than for welding similar plate

Procedure

1. Set up the plates as for welding, unit 19, except that the tacks are made with bronze alloy.
2. Adjust the flame.
3. Proceed with brazing in much the same manner as when welding.
4. Observe the tendency of the bronze to tin the upstanding leg of the joint over large area.
5. Correct the flame angle and the distance of the inner cone from the work until this excessive tinning is overcome. Figure 32-3 shows the correct flame angle.
6. Bend the finished joint in the manner used to test the welded fillet joint. Examine the root of the joint for uniformity of bond.
7. Braze the opposite side of the joint. Note the difficulty in achieving a good bond on the upstanding leg.

Brazed Tee Joints

Fig. 32-2 Welder's eye view   Fig. 32-3 Brazed fillet joint

REVIEW QUESTIONS

1. What factors make it difficult to obtain a good bond on the second side of the joint?

2. What preparation is necessary to obtain a good bond on the second side of the joint?

3. How does the flame for brazing the reverse side of the joint differ from that used on the first braze? Why is there a difference?

4. How should material be prepared for brazing?

5. What makes this joint more difficult to braze?

Unit 33

BRAZING BEVELED BUTT JOINTS ON HEAVY STEEL PLATE

The lower temperatures used for brazing as contrasted with fusion welding make this process good for many jobs. Melting of the base metal is avoided, extensive preheating is not necessary, and expansion and contraction are not as severe problems as they are in fusion welding.

The student should acquire an understanding of brazing and will develop skill in this process. This unit provides an opportunity for brazing heavier steel than that used in previous units.

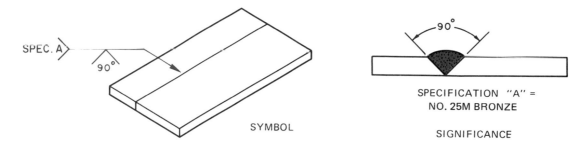

Fig. 33-1 Beveled butt joint

Materials

Two pieces of 1/4-inch thick steel plate, 2 in. X 6 in. each
1/8-inch diameter bronze rod or coated bronze rod
Flux
Welding tip one size larger than for welding similar plate

Procedure

1. Prepare the plate edges so that they make an angle of 90 degrees when they are brought together, figure 33-1.

   Note: If the plates are flame-cut, lightly grind or file the cut edges until all oxides are removed.

2. Align and tack the plates as for welding, figure 33-2.

3. Braze the joint. Be sure the plate edges are tinned to the root of the joint. Be careful not to overheat the tip edges of the plate as the brazing proceeds.

4. Make more joints of this type but apply the alloy in two layers. Apply the second layer with a weaving motion, alternating the rod and flame in the same manner as in welding heavy plate. When making a multiple-pass braze, be sure that each bead, except the finished bead, is slightly concave and that the alloy tins well up on the sides of the joint, figure 33-3.

5. Vary the angle of the flame and observe the results.

Brazing Beveled Butt Joints on Heavy Steel Plate

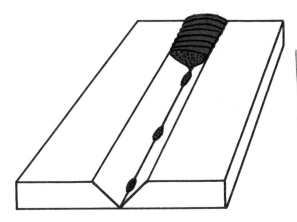

Fig. 33-2 Single-pass braze

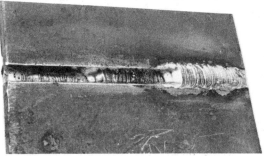

Fig. 33-3 Multiple-pass braze

REVIEW QUESTIONS

1. What effect does the flame angle have on the penetration into the root of the groove?

2. What effect does the flame angle have on the appearance of the finished bead?

3. Why is the included angle for this joint 90 degrees instead of the 60 degrees used when welding a beveled butt joint?

4. Should the surface of the beads be flat, convex or concave?

5. Should the bottom sides of the material being brazed be cleaned? Why?

Unit 34

BUILDING-UP ON CAST IRON

Worn areas of cast iron machine parts may be built up with bronze by the brazing process. The new surface may be machined if necessary. The resulting job is often as good as a new part.

Brazing on cast iron has advantages over fusion welding, largely because of the lower temperatures used. Brazing saves time, uses less gas, and involves less expansion and contraction of the metal. This unit gives practice in this process which is often called *bronze surfacing*.

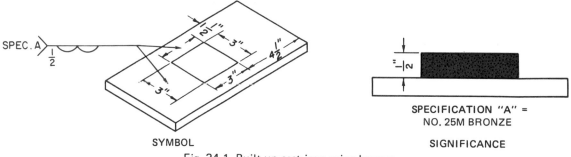

Fig. 34-1 Built-up cast iron using bronze

Materials

Cast iron
1/8-inch diameter bronze rod or coated bronze rod
Suitable flux
Welding tip one size larger than that indicated for welding metal of similar thickness

Procedure

1. Clean the surface to be brazed. Remove all rust, scale, and oil.
2. Preheat the cast iron with the flame before attempting to braze.
3. Build up an area about 3 inches square with the bronze rod using beads 3/4 inch wide, figure 34-1. Make sure that each bead is fused into the preceding bead. Use a standard dry flux for this operation.

    Note: If a Hi-bond® cast iron brazing flux or its equivalent is available, use this in with the standard flux. Note the ease of tinning on the cast iron surface.

4. Make the first layer about 1/8 inch high. Apply a second bead over the first as soon as the work has been inspected and before the plate has a chance to cool.
5. Apply a third and fourth bead. Inspect each bead and correct any faults when making the next bead.
6. Grind one edge of a piece of cast iron 1/4 inch thick and 6 inches long until it is square and clean. Stand this plate on edge so that a bead can be applied to the ground edge, figure 34-2.

Building-up on Cast Iron

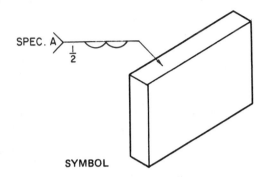
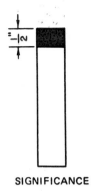

Fig. 34-2 Building-up a ground edge

7. Apply a bead along this edge, moving the flame and rod so that the bronze flows to each side of the plate but does not overhang.

8. Inspect the finished bead for uniformity of thickness and ripples. Note the presence or absence of overhang.

9. Deposit additional beads over the first until the edge is built up to a height of 1/2 inch above the original plate, figure 34-2. Check for uniformity.

REVIEW QUESTIONS

1. How do the tinning characteristics of cast iron compare with those of the same thickness of steel?

2. Why is preheating of cast iron necessary?

3. Draw a sketch of the cross section of the bead and casting in step 8.

4. Can cast iron be too hot to braze? What happens?

5. Is there an advantage to using stringer beads in brazing instead of wide beads? Why?

87

Unit 35

BRAZING BEVELED JOINTS ON CAST IRON

The quality of a brazed joint on cast iron is greatly affected by the surface preparation. The surfaces must be clean to allow good tinning and to get a strong bond. The two surfaces to be joined must have enough area to provide a strong joint.

Testing the brazed joint to destruction demonstrates some of the qualities of a good joint. This unit provides an opportunity to make and test brazed joints on cast iron.

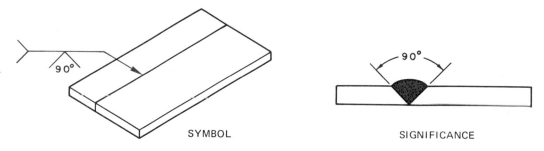

Fig. 35-1 Beveled butt joint

Materials

Two pieces of cast iron 1/4 inch or thicker, 2 in. X 6 in. each
1/8-inch diameter bronze rod or coated bronze rod
Suitable flux

Procedure

1. Grind or machine one edge of each of the two pieces of cast iron to 45 degrees. When the two pieces are aligned, the included angle of this opening should be 90 degrees, figure 35-1.

   Note: The included angle for brazing is normally made larger than that for welding the same size material. The reason is that in a bonding process such as brazing, the wider V presents more bonding surface for the brazing alloy; as a result a stronger joint is formed.

2. Align the pieces and preheat them by playing the flame over the joint until the work becomes dull red in color.

3. Sprinkle some Hi-Bond® flux along the joint and tack each end of the assembly.

4. Braze in a manner similar to that in unit 33. Be sure that good tinning action takes place along the sides of the V. Do not try to complete the braze in one pass, figure 35-2. Attempts to build up too great a thickness in one step usually result in poor tinning and a weak joint.

5. After inspecting the braze for appearance, cut the piece in two and place each piece in a vise with the brazed joint slightly above the vise jaws. Bend one piece toward the root of the braze and one toward the face of the braze, figure 35-3.

Brazing Beveled Joints on Cast Iron

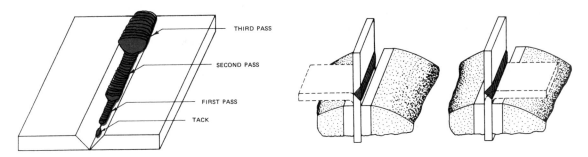

Fig. 35-2 Multilayer brazed joint   Fig. 35-3 Testing the braze

6. After these pieces break, inspect the break for a good bond. This is indicated by a coating of bronze on the cast iron and small particles of cast iron clinging to the bronze.

7. Prepare two more pieces. After beveling, draw a coarse file over the beveled surfaces to roughen them. This presents a surface of greater area to the brazing alloy. In addition, any free graphite which may be left on the beveled surface by the grinding operation is removed.

8. Braze this joint and test as before. Compare the bond strength of these plates with the first set tested.

REVIEW QUESTIONS

1. What prebrazing operation is most important to insure success in making a brazed joint?

2. What is the effect if brazing is attempted on a surface coated with free graphite?

3. When the joints made in steps 4 and 7 are tested, is there any difference in the amount of bending necessary to break the joints? Which joint requires the most bending? Why?

4. What color should the base metal have for good brazing?

5. Why is the included angle of the brazed bevel joint wider than for welding?

Unit 36

SILVER SOLDERING NONFERROUS METALS

Silver soldering or *silver brazing* is, in reality, low-temperature brazing. A typical alloy, such as the one used in this unit, contains 80 percent copper, 15 percent silver, and 5 percent phosphorus. This type of alloy is effective at temperatures well below the melting point of brass and copper, the metals most commonly silver-soldered.

A number of different alloys with various melting points and fluxes is available.

The development of skill in silver soldering greatly increases the scope of the work which a welder can undertake.

Materials

Two pieces of strip brass or copper 1/16-inch thick, 1 in. X 6 in.
Two pieces of brass or copper tubing, one of which just slips inside the other.
Handy-Flux® or equivalent
Sil-Fos® brazing alloy or equivalent, 1/16 in. X 1/8 in. X 14 in.
Welding tip, see chart 6-2 for size.

Procedure

1. Prepare the strips to be brazed by wire brushing, rubbing with emery cloth or steel wool, or by dipping in an acid bath to insure absolute cleanliness.

    Note: All welding and brazing operations are more successful if attention is given to surface cleanliness. In the silver brazing operations, failure to observe the proper precautions results in joints of low strength and poor appearance. A bright surface does not necessarily mean a clean surface from a welding or brazing viewpoint. Surface oxides sometimes appear very bright.

2. Apply a thin layer of flux to both surfaces of the strips that are to make contact in the braze. Allow this fluxed area to extend somewhat beyond the area to be brazed.

3. Set up the work so that the lapped surfaces to be brazed are in close contact, with the ends lapped 1 inch.

    Note: Proper joint spacing is very important in silver brazing and silver soldering. The ideal clearance between the surfaces to be brazed is 1 1/2 thousandths of an inch (.0015 inch). If this clearance is kept, the joint develops its maximum strength. The strength of the finished joint falls off rapidly as the clearance is increased beyond .0015 inch. At the same time, the amount of expensive silver soldering alloy used rises at a rapid rate.

4. Apply a 2X or 3X flame to the work with a back-and-forth motion to heat the brazing area. Do not apply the flame directly to the brazing rod.

5. Heat the work until the alloy can be applied to the joint a short distance from the flame. If the work is hot enough, the heat is conducted into the alloy causing it to melt and flow into the joint by capillary action.

6. During the brazing operation, the flux serves as a temperature guide, as well as a cleaning agent.

   Note: When the heat is first applied, the flux dries and turns white. As the amount of heat in the work increases, this white powder starts to melt, forming small beads of molten flux on the surface. Further heating causes the flux to become more fluid and flow out over the work surface in a thin, even coating. When this condition is noted, the work is at the proper brazing temperature, 1,300 degrees F. Temperatures beyond this point tend to make the molten flux *crawl* or leave bare areas on the work surfaces which oxidize. The result is a poor bond and unacceptable appearance.

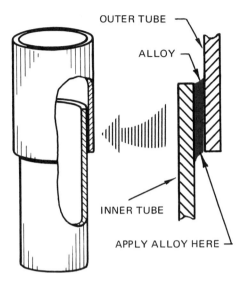

Fig. 36-1 Silver brazing tubes

7. Test the finished joint by trying to peel one of the lapped pieces from the other. If the braze has been properly made, the metal tears before the joint breaks.

8. Clean two pieces of copper or brass tubing inside and out. Make sure that one piece just slips inside the other, as in figure 36-1.

9. Place flux on both pieces and slide one into the other for a distance of 1 inch. Set this assembly on the bench or in a vise in a vertical position with the larger tube at the top.

10. Proceed with the brazing, making sure that all precautions outlined above are observed. Apply the alloy only on the outside of the joint. Make sure that the heating is complete.

11. Cool the finished work and note the thin, even line of alloy around the tube. Look into the tube and note that the alloy flowed by capillary action up between the two tubes and formed a small bead at the end of the inner tube. A further check can be made by hacksawing the joint diagonally and noting the white silvery appearance of the alloy between the entire lapped surfaces.

REVIEW QUESTIONS

1. If the base metal appears highly oxidized after the braze is completed, what factors must be checked before this condition can be corrected?

Basic Oxyacetylene Welding

2. What are the factors that must be observed to produce good silver brazed joints?

3. What is a 3X flame?

4. What alloys are contained in typical silver solder?

5. How can material be prepared for silver soldering?

Unit 37

SILVER SOLDERING FERROUS AND NONFERROUS METALS

Silver soldering, silver brazing, and low-temperature brazing are terms used for the same process in industry. To be technically correct, the process done at a temperature greater than 800 degrees F. is called *brazing*. The melting points for the different brazing alloys are:

Easy-Flo®	1,175° F.
Sil-Fos®	1,300° F.
Phos-Copper®	1,600° F.
Bronze	1,750° - 1,800° F.

The operations in this unit are done at a lower temperature than in the previous unit. The Easy-Flo® alloy is used with either ferrous or nonferrous metals.

All silver soldering or silver brazing operations require very small amounts of the alloy. Any attempt to use these alloys as welding rods or high-temperature brazing rods results in a weak and costly joint.

Materials

Strips of steel, stainless steel, copper, and brass about 1/16-inch thick, 1 in. X 4 in. each
Handy-Flux®
Regular Easy-Flo® and #3 Easy-Flo® 1/16-inch diameter alloy
Welding tip, see chart 6-2 for size.

Procedure

1. Prepare steel plates following the procedure of unit 36.
2. Set up two of the plates, lapping the ends about 1 inch to make an assembly 1 inch wide X 7 inches long. Be sure that the opposite end of the top plate is supported so that the two plates are kept parallel.
3. Braze the joint, using the same flame adjustment and technique as in unit 36. Apply the alloy (regular Easy-Flo®) by wiping the rod slowly on the one-inch face of the lap on the top side only. Apply enough heat to cause the alloy to flow between the lapped surfaces completely.
4. Cool the brazed joint and observe the opposite face of the lap and both edges. The alloy should be visible all around.
5. Try peeling one lapped plate from the other. This is impossible if the joint is properly made.
6. Set up two stainless steel plates and repeat the steps above.

   Note: The upper limit of temperature is very critical when brazing stainless steel. Overheating causes the formation of chromium oxide which can be removed only by filing or grinding. Flux does not remove chromium oxide.

Basic Oxyacetylene Welding

7. Experiment with stainless steel joints by heating two of the plates until the color indicates that oxide has formed. Then proceed with the brazing as in previous joints and observe the results.

8. Check the joint of step 7 visually for complete bonding; peel one plate from the other.

    Note: Tensile strength tests made on properly brazed joints on stainless steel break outside the brazed area at a load of 100,000 to 120,000 pounds per square inch. The actual strength of the braze is somewhat higher than the above figures.

9. Set up two brass plates to make a fillet- or tee-type joint. Make sure both sides of the joint are covered with flux.

10. Braze this joint using #3 Easy-Flo®. Note that this alloy does not flow as freely as other types, but allows a slight fillet to build up. This is desirable in certain applications.

11. Cool and inspect the finished braze. In particular, check the draw through of the alloy on the reverse side of the joint.

12. Hammer the upstanding leg flat against the bottom plate. Note that the color of the alloy is very close to the color of the work. This is desirable in many jobs where color match is important.

13. Prepare two more plates but, in this case, make a butt joint. Test in the same manner as for a butt weld and observe the results.

CONCLUSION

1. Thoroughly clean the joint area.
2. Apply flux to the joint area.
3. Secure parts in proper alignment with necessary clearance (.0015 inch to .003 inch).
4. Use large soft carburizing flame held 1 inch to 3 inches from the work.
5. Heat the parts until the flux liquefies (dull red on ferrous materials).
6. Apply the alloy, keeping the torch in constant motion until the alloy flows completely through the joint, leaving a smooth fillet on each side.
7. Avoid overheating and remelting.
8. Cool slowly and remove flux residues with hot water.
9. Specific information can be obtained from alloy manufacturers.
10. This technique can be used to join ferrous and nonferrous metals including stainless steels used in hospitals and food processing, high pressure piping, and carbide tool tipping.

REVIEW QUESTIONS

1. If the joint is overheated when low-temperature brazing stainless steel, what procedure must be followed to obtain a satifactory braze?

2. Is it possible to make fillets when using silver brazing alloys? Explain.

3. What is the proper clearance between the lapping surfaces for maximum joint strength?

4. How do butt-brazed joints compare with lapped joints?

5. What can be said about the amount of filler metal required for a given joint?

INDEX

Acetylene
   chemical properties, 7–8
   generation of, 7
Air, gases in, 2

Backhand, 70–71
   butt joint, heavy steel plate, 70
   procedure, 70–71
   welder's eye view, right-handed, 70
Beads, running with bronze rod
   diagram, 76–77
   filler metal and flame, contact of with work, 77
   procedure, 76
Beads, running and observing
   diagram, 41
   flame, motion of, 43
   flame, types of, 43
   manipulation of torch by right-handed operator, 42
   penetration of, 41, 43
   procedure, 41, 43–44
   running of, example, 42
   steel, reactions of, to heat, 43
   test weld, bending of, 42, 43, 44
   width control, 41
Bevel cutting, 31–32
   manual, 31
   procedure, 31–32
   tip alignment, 31
Beveled butt joints, backhand welding
   cross-section, 72
   diagram, 72
   procedure, 72
   rod and torch motion, 73
   welder's eye view, 73
Beveled butt joints, brazing on heavy steel plate
   diagram, 84
   multilayer, 84, 85
   procedure, 84
   single-pass, 85
Brazed joints
   fillet joint, 82
   procedure, 82
   tee, 82
   torch and rod motion, 83
   welder's eye view, 83
Brazing with bronze rod
   cleaning of metals, 74

flux, 74, 75
nature, 74
procedure, 74–75
zinc oxide, 75
Butt welds, light steel plate. *See also* Heavy steel plate
   diagrams, 48, 49
   penetration, 49
   procedure, 48–49
   on 16-gage steel, 49

Cast iron, brazing beveled joints on
   diagram, 88
   included angles of, 88
   multilayer, 89
   procedure, 88
   roughening of surfaces for, 89
   testing, 89
Cast iron, building up worn surfaces on
   bronze, deposition of, 86
   diagram, 86
   ground edge, 87
   procedure, 86–87
Check valves, in hose connections, 21
Cutting with flame
   advantages of, for mild steel, 24
   combination torch for, 24
   comparison chart for tip sizes, 25
   equipment, 24
   gas pressure vs. kerf widths, 26
   hazards, 26
   machine torch, 25
   procedure, 24
   shape-cutting torch, 25
Cylinders
   acetylene, 5
   oxygen, 4–5
   valve for, 5

Discs, making of, 34

Excessive pressure in hoses, 22

Fillet welds, light steel plate
   descaling, 57
   making, flame motion for, 57
   procedure, 56, 57
   tee joint, 56

Index

Fillet welds *(continued)*
  tee weld, 56
  testing, 57
  undercutting, 57
  welder's eye view, 57
Flame, acetylene
  acetylene feathers, 18
  carburizing, 17, 18
  neutral, 17
  oxidizing, 17, 18
Flame, adjustment of, 22
Flat corner welds, 51–53
  corner, flame on, 52
  corner joint, 51
  outside corner weld, 51
  penetration, check on, 53
  procedure, 51, 52
  testing, 53
  welder's eye view, 52
Fluxes, 2, 74, 75, 91

Gages
  diagram, 10
  operation of, 9, 11

Heavy steel plate
  bead on, 59
  beveled butt weld, 68–69
    diagram, 68
    example, 68
    procedure, 68–69
  butt welds on
    diagram, 62
    example, 63
    open square type, 62
    procedure, 62
  fillet welds
    diagram, 66
    example, 67
    procedure, 66
    welder's eye view, 67
  lap welds on
    diagram, 64
    example, 65
    procedure, 64
    testing, 65
    welder's eye view, 65
  multilayer, 69
    procedure, 69
  rod, manipulation of
    diagram, 61
    procedure, 61
  tip size and gas consumption, 60
Holes, cutting of
  diamond point chisel to start, 33
  enlarging of hole, 34
  large, 34
  preheating, 34
  procedure, 33, 34
  sequence, 33
  starting of cut, 34
Hose connections, threading of, 21

Kerfs. *See* Cutting with flame

Lap joints, brazed
  diagram, 80
  procedure, 80–81
  rod and torch motion, 81
  tinning action of alloy, 80
  welder's eye view, 81
Lap joints, light steel
  lap joint, 54
  lap weld, diagram, 54
  making of, 55
  procedure, 54–55
  testing, 55
  uneven heating, 54
  welder's eye view, 55

Oxyacetylene welding procedure
  advantages vs. disadvantages, 1–2
  equipment, 1
  eyes, protection of, 2
  flame and sparks, control of, 2
  fluxes, fumes from, 2
  hazards, 1
Oxyfuel welding and cutting equipment, 20
Oxygen
  chemical properties, 7

Piercing. *See* Holes, cutting of

Regulators
  acetylene, example, 10
  diagram, 9
  discussion, 9
  oxygen, 10
  threads of, 20
Rods, welding, making beads with
  deposition rate, 46
  diagram, 45
  dipping motion of rod and flame, 46

Index

fixed, 46
procedure, 45, 46
use, 45

Scale, 57
Setup of equipment, 20, 21, 22
Shutdown and purge procedure, 22
Silver soldering, of both ferrous and nonferrous metals
    alloys, melting points of, 93
    amounts to use, 93
    procedure, 93–94
    stainless steel, 93
    summary, 94
    tensile strength, 94
    testing, 94
Silver soldering, nonferrous metals only
    alloy for, 90
    brazing tubes, 91
    cleaning of strips, 90
    flame, use, 90, 91
    flux, 91
    joint spacing, 90
    testing, 91
Slag, 29
Square butt brazing on light steel plate
    diagram, 78
    procedure, 78
Straight lines, cutting of
    cutting of sheet when cherry-red hot, 28
    faults, common, 30
    machine-cut, 29
    manual-cut, 29
    procedure, 28, 29
    slag, 29
Symbols for welding
    all-around, 39
    chain-intermittent, 38, 39
    examples, interpretations of, 37, 38
    field weld, 39, 40
    fillet, 39, 40
    root opening vs. penetration depth, 37
    staggered intermittent, 38, 39
    standard, 36

Tack welds, on light steel plate, 48, 49
Tee welds. *See* Fillet welds, light steel plate
Temperature scale, 7
Tips, for welding torches
    brands, chart, 15
    care, 14, 15
    pressures for, 14
    size selection, 14
    size vs. material gage, 14, 15
Torches, welding
    injector type, 12
    medium pressure, 13
    mixer in, 12
    mixing action in, 12
    views of, 12, 13

Wrenches, use in setup, 21